JANUA LINGUARUM

STUDIA MEMORIAE
NICOLAI VAN WIJK DEDICATA

edenda curat

C. H. VAN SCHOONEVELD

Indiana University

Series Minor, 151

OPPOSITIONS IN MORPHOLOGY

As Exemplified in the English Tense System

by

IRINA KHLEBNIKOVA

1973

MOUTON

THE HAGUE · PARIS

Translated from the Russian

LIBRARY OF CONGRESS CATALOG CARD NUMBER: 72-94476

Printed in Belgium by NICI, Ghent

PREFACE

In connection with the widespread use during the last two decades of methods and individual ways of linguistic description, somehow linked with transformational (generative) grammar, there arises the danger of neglecting the obvious fact that no matter how varied are the relations among separate units of the language (especially those among bilateral units having both form and meaning), the basis of any language is a definite system, a paradigmatic framework. The cells of this framework are filled with units natural for a particular language, which stand in interdependent, paradigmatic relations to one another.

The main object of transformational *(T)* grammar is units bigger than words, first of all the sentence, and more rarely a word-combination; describing sentences as trees, establishing *T* dependences among them — the basis of which is more logical than linguistic or grammatical — analysing the syntactic and semantic components of the sentence, all these methods to a great extent ignore the word as a unit of grammar and its forms, which constitute the grammatical basis of the language. The word is the ultimate constituent of the sentence, its initial element, and it is the functioning of its various forms on the syntagmatic axis and their mutual distribution that create a word-combination, a sentence, all that is called connected speech.

In the sphere of the sentence (and word-combination) *T* grammar does not designate as its task the creation and description of systems of sentence-patterns in their mutual correlation, so that varied two-member and one-member, standard and non-standard structures of the sentence fall out from syntactic descriptions, as do all the varied types of composite sentences in relations among themselves in the system.

In other words, *T* grammar has not produced any PARADIGMATICS of morphological forms of the noun and verb (and their subclasses) which would express grammatical categories; neither are the grammatical categories themselves defined — the plane of content of morphological forms. The inventory of types and models of sentences and word-combinations of any language has not been produced by *T* grammar either.

Hence the absence of such notions as morphology and syntax in many recent works in the field of grammar.

In light of the above, the author of this work deemed it her task to partially fill in the gap formed in the presentation of language as a system of interconnected units, and to show, on the basis of the theory of oppositions born in the depths of the Prague school of linguistics as a development of the ideas of F. de Saussure, the associative paradigmatic relations among morphological units in the system, in particular, among categorial forms of the English verb. The author's deep conviction is that the description of associative relations of morphological units in the system should precede the description of their co-occurrence in the linear chain, that is in the sentence, in an utterance of any length.

The oppositional analysis is a fruitful method of research and description of various properties natural for the elements of language structure and, first of all, in the field of morphology, since oppositions imply connection, and connection is an obligatory component of relations among any units of language.

However, not always are oppositions discovered by linguists where they actually exist; there are certain aspects of language where they are hardly possible. Up to now it has been impossible to consider as completely elaborated the following: the notion itself of the opposition (and contrast), the criteria of detecting oppositions, their inner structure, the sphere of their action at different levels of language, and many other related facts.

It is the elucidation of a number of questions concerning oppositions generally and in morphology, in particular, to which this publication is devoted.

The theory of oppositions would have been a dead scheme had its action not been demonstrated on language material. Every theory gains if it is supported by facts, and the actual language material (at the present stage of language science development) should also be considered on the basis of the general theory of the given science. That is why oppositions are shown by the English verb, which, for the purpose, possesses favourable qualities.

The application of the method of oppositional analysis to concrete language material makes it possible, due to the objective and formal nature of the method, to remove some essential differences in the interpretation of the content of morphological units which are subjected to investigation and comprehension in their interrelation.

The limitation of the volume of this publication prevents us from considering here a number of adjacent questions concerning morphological units, such as their classes and subclasses, morphemic structure, homonymy, synonymy, variation, their use in speech, etc. All this should be the object of a special discussion. Here we shall confine ourselves to that side of morphological units in which oppositions are directly active, viz., their paradigmatic grammatical meaning.

As to the search of *T* relations among separate language units or generating and generated structures, they, we think, present one of the tasks of describing complex syntactic units of language, which, precisely due to their complex nature, present a wide field for logical, structural, and semantic research. This is a stage of a more subtle linguistic description to be based on a previously established system of forms and models of a particular language in their paradigmatic correlations.

TABLE OF CONTENTS

ABBREVIATIONS

NOTATIONS

C — continuous
DF — distinctive feature(s)
F I — future I (simple future)
F II — future II (future-in-the-past)
I — indefinite
IM — invariant meaning
MS — microsystem
P — perfect
PC — perfect-continuous
Pr. — present
Pt. — past (preterit)
SM — subjunctive mood
VF — verbal form(s)

1.

INTRODUCTION

The qualitative change in linguistic research methods, which became especially pronounced half a century ago after the ideas of Ferdinand de Saussure (and those of J. Baudouin de Courtenay in Russia) became the property of broad circles of linguists the world over, immediately suggested a whole complex of problems which had found no reflection in pre-structural (or pre-Saussurian) linguistics. There appeared new notions and terms to denote them, and the point of view from which language material was examined also gradually changed.

It is hardly possible now, in the second half of the twentieth century, to remain exclusively on positions and conceptions which are more than half a century old, the main feature of which was lack of any consistently applied method of treatment and description of language material, and which formed the basis for the linguist's method of work with language data of any kind.

However, the whole wealth of information on different aspects of language obtained by pre-structural linguistics is widely used in the practice of structural research, and it is not necessary to separate all that was done earlier from today's structural investigation of different trends.

Still, if we analyse or synthesize any set of language units, such an analysis should be done using everything positive accumulated by structural linguistics, since otherwise it would mean not progress but regress, neglecting the basic principle of advancing forward —

responding to everything newly discovered by science, because newness in science, in the final count, is always progressive, though it is not always equally well interpreted through different conceptions.

The development of linguistics has, on the whole, been dialectical; since war was proclaimed on the atomistic approach to the analysis of language facts a trend was clearly taking shape to know the connection between different phenomena, and to discover cross interdependences within different language units and their correlations, both associative, paradigmatic, and linear, syntagmatic.

Methods of linguistic description, which arose in the depths of various schools of structural linguistics, together and separately serve the purpose of studying the properties of language structure, the object of linguistics, and, hence, are aimed at establishing the nature of interconnection of its units.

The inevitability of the rise and tumultuous development of all structural methods and individual ways of studying the language material was conditioned by the whole course of world science and different branches of human knowledge. Such interdependence of language science and other fields is due, among other factors, to the introduction of the notion of 'structure' into all fields of science.[1]

The applicability of different methods depends on the concrete material and the tasks of investigation. Thus, if the aim of linguistic description is the content of a certain totality of morphological units, then the most productive will be oppositional analysis, since the very nature of categorial relations in the system suggests such types of relations, contained within the nature of logical oppositions.

In the field of MORPHOLOGY — part of language and the division of grammar, which studies it — different methods are at work in the examination of the grammatical structure of the word.

[1] On the notion of structure in different sciences see *Notion* (1957). On relations within a structure see Russell (1948) part IV, ch. 3; Šaumjan (1962) 85; Šaumjan (1965) 11-14; Lomtev (1958) 6.

There are also such trends (descriptive linguistics, syntagmatic structuralism, transformational grammar) which do not accept the opposition of morphology and syntax for different reasons, the explanation of which, as well as that of the conceptions of various schools of structuralism, is beyond the task of this study. Morphology, as an independent object of study, possessing its own ontology and, accordingly, methods, has been singled out as an independent sphere of grammar since time immemorial, but it is presented to-day with more or less clearly defined boundaries in the Prague school of structuralism, i.e. in functional grammar (which is also joined by the London and French schools), and in the works of a number of Soviet linguists.

By morphology we shall understand that part of grammar (as of a science, which studies rules of the structural organization of the meaningful elements of a language) which concerns itself with extracting the system of paradigms from a mass of speech events, the paradigms being built on the principle of unification on the basis of categorial[2] signs common to each of them, serving as the basis for comparison of the members of each paradigm among themselves. Thus, if in a certain language a set of formtypes unites in a paradigm, having in common the expressions of meaningful relations among words as parts of the sentence (i.e. the grammatical category of case), then every separate formtype would denote a given concrete relation (that of subject, object, etc.), i.e. a separate case.

Morphological forms represent a bilateral sign, having its plane of expression (form, material cover) and plane of contents (meaning, through which the given unit correlates with objects and relations of objective reality). Hence, morphology concerns itself with the structure of a categorial form (grammeme)[3] and the

[2] The term 'categorial' (categorical) was introduced into Russian linguistics by Smirnickij (1957) 30. The term is used nowadays by many linguists outside the Soviet Union, for instance Kurylowicz (1956) 1; Zawadowski (1959) 103; Hartmann (1962) 456-458; Joos (1964) 14.

[3] In order to avoid the polysemantic term 'form' we occasionally designate a categorial form as a unit of grammatical information by the term 'grammeme'. This term is parallel to other emic denominations, such as 'phoneme', 'mor-

meaning it contains and due to which it becomes a member of a definite regular series. All this comes into the sphere of PARADIGMATIC MORPHOLOGY which may be contrasted to SYNTAGMATIC MORPHOLOGY[4] the task of which involves studying the use of every paradigmatic form in the real text or speech, where its categorial meaning is realized.

In the sphere of syntagmatic morphology, since it deals with the distribution of environment of categorial forms in linear order, other ways (methods) of analysis (outside the oppositional one) are effective, which correspond to the object, first of all the distributional method as well as those of Immediate Constituents and Transformations.

At the present stage of linguistic research, when a great amount of actual language material has been accumulated of the most widely known Indo-European languages, it would be expedient, before undertaking a syntagmatic analysis of the instances of the use of paradigmatic forms in context, to define the morphological system itself in all its variety and in all the complexity of the ties and relations of its units, using data accumulated by grammar and postulating the nature of the relations within the system by introducing into the analysis some elements of hypothesis, which is further verified on actual material.

Thus, it is absolutely unnecessary, prior to explaining relations in the Russian case system, to collect examples of all the uses of cases widely known in grammar. But correctly ascertained paradigmatic systemic connections would reveal the essence of the

pheme', and it denotes, in the present work, a unit which comes next to the morpheme on the scale of language levels according to the complexity of its structure. This unit represents an indivisible complex of grammatical (categorial) meaning, such as, e.g., the past indefinite or the present perfect, independent of the number of its separate constituents.

On emic denominations see Akhmanova (1961) 115-121.

The term 'grammeme' is given various interpretations by linguists, sometimes even by one and the same author. Pike (1954-1960) part I and II uses the term 'grammeme', which is substituted in part III, 1 by the term 'tagmeme'. This change is explained in Pike (1958) 273-278.

4 The notion of 'syntagmatic morphology' is introduced in the articles: Khlebnikova (1965a) 127; Trnka (1966) 165-169.

function of cases in speech and would help to give them meaning as links in the general chain of relations, connecting units of language among themselves. Hence, at the present stage of grammar research it is more expedient first to define the system and then check its actualization in speech.[5]

If the definition of the inventory proper of categorial forms does not present any great difficulties, then defining the relative content of each wordform is not a simple matter and calls for wide generalizations and calculation of all possible directions into which the connections of every concrete form are spreading. This task proved insurmountable for many grammarians of various schools, and in most cases the interpretation of the content of forms was permeated with the subjective judgments of researchers. Some linguists did not set themselves such a task at all. Thus the representatives of descriptive linguistics did not concern themselves with extracting the system of categorial relations, but dwelt mainly on the classification of morphemes or morphs (on the basis of commutation) by similarity or distinction. Hence, descriptive linguistics did not work out a method for analysing categorial relations in a paradigmatic system of those parts of speech (classes of words) which possess such a system. Transformational grammar has no such method either.[6]

One may state that a certain grammatical (morphological) system has been established correctly if all the notional categories and the forms expressing them are arranged into a framework according to mutual correlations among the units of the given system (or microsystem), and if not a single form (and its meaning) remains isolated outside the system.

The working out of special methods for analysing categorial meanings of morphological paradigmatics started, mainly, in the Prague school and was reduced to the method of oppositions. Oppositions became a clearly enough defined method of paradigmatic analysis of categorial, systemic forms, and is often applied to various units of linguistic structure. Still it is a comparatively

[5] The same opinion is expressed by Bull (1960) 34.
[6] On paradigmatics in transformational grammar see Šaumjan (1965).

new method and is far from clear. It is high time, however, to specify how and on what basis we should define the plane of content of morphological units and what criteria of objectivity may be set up.

Since the theory of oppositions in morphology, to be discussed here, may be graphically illustrated on concrete language material only, we shall take as such material the tense-aspect system of the English verb as the most differentiated one among the systems of the Indo-European languages. Such a system will be the best illustration of the productiveness of oppositional analysis as a method penetrating the nature of associative ties in morphology and, hence, presenting the most formalized way of linguistic analysis of the plane of content of morphological units (grammemes) which, as any unit of semantics, is difficult to formalize.

2.

THE REASONS FOR THE RISE OF THE NOTION OF OPPOSITION AND THE METHOD OF OPPOSITIONAL ANALYSIS. OPPOSITIONS AS A MANIFESTATION OF THE PLANE OF CONTENT OF MORPHOLOGICAL FORMS

The inner quality of language, as a complex structure made up of a plurality of systems (and subsystems) is the opposition of units. Contrasting was discovered when, due to the accumulation of a certain sum of actual language data, linguists gradually came to understand the fact that language is not a chance conglomeration of isolated words and their forms, but is a system. The very concept of 'system' meant interconnection, joining on the basis of community and distinction, a totality of related elements. That is why back in 1916 in Ferdinand de Saussure's '*Cours*', one can find many assertions to the effect that language is in itself a system: "La langue est une système dont toutes les parties peuvent et doivent être considérées dans leur solidarité synchronique",[1] and "Le mécanisme linguistique roule tout entier sur les identités et des differences, celles-ci n'étant que la contrepartie de celles-là".[2]

These words reflected a turning point in the development of linguistics as a science, and scientists of different countries, influenced by de Saussure or independently, came by different paths to an analogous opinion, since such was the trend of movement in the science of language, a science which could not stay aloof of the general course of development of all sciences and which

[1] Saussure (1915) 124.

[2] Saussure (1915) 151. Similar views are propounded by many followers of de Saussure, e.g. Frei (1929) 42.

increasingly concentrated its attention on the knowledge of the properties of its object — language and its separate units.

The very term 'system' implies a definite order of its elements, and the existence of regular ties between them. Any order implies a certain symmetry, inner organization, succession, and interconnection. All these qualities are characteristic not only of the language as a whole, but of its various sections — microsystems. It is natural, therefore, that linguists are increasingly turning to oppositions, which graphically demonstrate a regularity in the organization of language elements and upon which relations in the system are built. A synchronic system, as L. Hjelmslev points out, is not static, since this system is language in action, a play of contrasts ("Le jeu des oppositions entre signes").[3] Contrasts within language units are increasingly attracting the attention of Soviet linguists. Thus, A. I. Smirnickij wrote in connection with grammatical category: "Any grammatical category should necessarily be represented by at least two categorial forms".[4] L. R. Zinder thus characterizes this thesis: "The presence or absence of a grammatical category means the presence or absence of corresponding contrasts. A grammatical category is a totality of language events contrasted in meaning and form (linked among themselves)"[5]. "Grammatical units are correlated to each other through a system of relations, which are reduced to semantic–functional contrasts" ... says V. G. Admony.[6]

The discovery of regular contrast as the main feature of the paradigmatics of any linguistic sign naturally led to the search for oppositions, which on the one hand are a method of investigating the nature of contrasting, and on the other, reflect the very essence of real relations in the system, i.e., their ontology.

Despite the versatility of the language, heterogeneity of its

[3] Hjelmslev (1928) 49 ff. In Jakobson and Halle (1956) 4-5 the authors draw attention to the fact that "Opposition and contrast are two different manifestations of the polarity principle." A polar analysis at all levels of language is exploited in Juilland (1961) 48.

[4] Smirnickij (1959) 8. Cf. Stepanov (1966) 123 ff.

[5] Zinder (1962) 122.

[6] Admoni (1964) 12.

elements, and the coexistence in the synchronic structure of layers of various epochs, the organization of the language components possesses a definite balance and simplicity, expressed, by the way, in that members of contrasts in microsystems conform to a symmetrical framework and reveal in their structural interrelations a recurrent model of organization of sets of separate forms, and of those ties contained within these units.

The simplest types of correlations, characteristic of contrasts of units belonging to different levels of language structure, are the relations of binarity (dichotomy), linking separate units through certain polar features. It is not without reason that the view is spreading in modern linguistics that analysis of polar relations should be the subject of study at all levels.

The very idea of contrast implies binarity of relations, so, no matter what resistance some linguists may show to the principle of binarity in studying language units, the very conception of a system and everything that follows from it frequently leads to the discovery of precisely these bilateral ties among language elements.

It is another matter that diversity is always present in correlations of different language units, and binarism does not mean homogeneity of relations at different levels of language; neither can we completely exclude the possibility of the existence of triple or some other connections for example. Still, close study of natural laws in the structure and correlations of linguistic units in speech and in the system, especially at the phonological and grammatical levels, leads to binarism as an initial type of correlation.

Are the conceptions and terms 'contrast' and 'opposition' synonymous? Certainly not. To say that some or other of the elements are contrasted does not mean to give an analysis of oppositions in a system or microsystem. Oppositions are a concrete manifestation of contrast, and not in all language units, at that; and if we aim at oppositional analysis of any section of the system, we should first of all give a definition of opposition and, the main thing, establish its nature, a problem which presents the most difficulty and the most interest. That is the task of this study.

Many relations of contrast are characterized by binarity though not all of them are oppositions. Binarity in the structure of different objects has been attracting the attention not only of linguists, but also of philosophers and logicians as well. Thus, B. Russell holds that any structure possesses some relations, and the simplest relations are most often binary ones.[7]

Argument still continues as to the applicability of the dichotomic principle to the analysis of relations between grammatical units joined on the basis of identity of some general distinctive feature *(DF)* and differentiation of some individual feature. Discussions of this principle in regard to phonology may be found, for example, in the article by A. A. Reformatskij "The dichotomic classification of distinctive features and a phonematic model of language",[8] and in regard to grammar in A. V. Isačenko's article "Binarity, privative oppositions and grammatical meanings".[9] The opinions of linguists diverge mainly in two directions: some consider that binarity is not a symbol of relations in a system, but a linguistic method to be conveniently used sometimes for investigating some object; others (R. O. Jakobson and A. V. Isačenko) believe that "analysis based on binary features helps with greater precision to determine the system of a given category, i.e. to reveal those general (invariant) meanings, which distinguish each member from the other co-members of a given category".[10] A. A. Reformatskij says that binarism is justified where it is not introduced from without, and in particular, where it is not imposed by logic, but is

[7] Russell (1948) 254 ff.

[8] Reformatskij (1961) 106-122.

Binarism is also discussed in Horalek (1964) 414-417; Budagov (1965) 247. Speaking of the Prague school, Martinet (1953) 585 writes: "The most favoured principle is the principle of binarity, according to which the whole language should be reducible to sets of binary oppositions. This principle should be valid ... in paradigmatic patterns, and on both the plane of meaningful units and that of phonology."

In search for some technique of analysis that would enable one to better explain the use of the verb units, Allen (1966) 24, citing the above words of Martinet, also comes to oppositions.

[9] Isačenko (1963) 39-56.

[10] Isačenko (1963) 39-40.

rooted in the very ontology of the language.[11] To proponents of the dichotomic principle belong those linguists who, somehow or other, are linked with the Prague school — R. Jakobson, B. Trnka, A. V. Isačenko, S. Karcevskij, N. S. Trubetskoy.

One cannot but agree with A. V. Isačenko, that "Correctness or inconsistency of any linguistic theory may be confirmed or rejected only by practice of further research, and not by references to the 'ontology' of the language or to the 'real language picture', since nobody has ever actually seen this 'real picture'. It follows from the above that the distinction between 'genuine' and 'euristic' binarism implies the presence of some oracle capable of distinguishing between relations 'really' existing in the language and those 'introduced' into the language from outside. Had such an oracle really existed, then all our science would have become absolutely superfluous. The only way to discover the 'language reality' is the way of scientific analysis".[12]

The fact seems to be indisputable that if grammatical (or phonological) phenomena did not show their systemic nature (revealed by all the preceding linguistic investigations), there would have been no need to look for the ways in which this systemic nature reveals itself, and to search for symmetry and a definite natural order in the organization of language units into a system; there would have been no sense in testing the language material for dichotomic division or classification. It is the recognition of systemic, and not chaotic nature in the organization of language units that called for putting forth those principles of classification which would best reflect this systematics and enable the establishment of the nature of interconnection among the units, and which would with maximum simplicity and precision reflect the natural laws of the system.

Much in nature and social phenomena and, hence, in the language, is built on binary contrasts. It is not by chance, therefore, that the universal nature of binarism is proclaimed by Ch. Ogden

11 Reformatskij (1961) 113-116.
12 Isačenko (1963) 44-45.

in his psycholinguistic monograph.[13] Much in the language is built around contrasts e.g. antonyms, classes of words with opposite meaning (abstract–concrete); the noun is opposed to the verb; the word is opposed to the word combination or the sentence; morphology is opposed to syntax, the complex sentence to the simple one, etc. However, all these cases are heterogeneous and cannot be indiscriminately reduced to oppositions.

Can we consider as an opposition such a clearly binary correlation as that of an attribute and the word modified by it, or the verb and its object, the verb and its subject? There can be no opposition in such cases as opposition implies the necessity of a common foundation for comparing its members, while we have no such foundation here, since every member of the sentence possesses a syntactic meaning of its own and enters binary relations in the speech chain relations which in their nature differ from oppositional ones in that they are subordinative and linear. Thus, in syntactic structures we deal with BINARY RELATIONS, and not with contrasts or oppositions.

The difference in the nature of relations among various language units provokes a desire, for example, to differentiate oppositions in morphology and contrast in syntax.[14]

First we may give the following criterion of distinction: we shall consider as opposition a correlation of two units on the paradigmatic axis, in a system or subsystem of grammemes; the paradigmatic axis of a language represents an extraction on the basis of meaningful associations of homogeneous elements, grouped according to the presence of some common categorial (distinctive) feature *(DF)*. Within the opposition the distinction

[13] Ogden (1932).

[14] Cf. Malmberg (1963) 6; Martinet (1960) 33.

Cf. the discussion of the meaning which the terms 'opposition' and 'contrast' acquire with linguists of different persuasions: Martinet (1964) 524 ff. Martinet calls the structure of relations existing among units of the same class 'paradigmatic structure', the "structure resulting from the relationships existing among the various classes — 'syntagmatic structure'." (p. 526).

A somewhat different idea of opposition versus contrast can be found with other linguists, e.g., Prieto (1954) 43-59.

between its two members lies in the individual manifestation of some particular *DF*. Thus, members of an opposition never stand 'nearby' in actual speech; they may be only substituted on the basis of commutation. For instance, *I am sitting*, *I was sitting*, *I shall be sitting* are different verb forms; they are not subordinated to each other and not juxtaposed in the speech chain, but denote different relations to the time of action as to an abstract grammatical category. In case these forms do find themselves side by side in an utterance, the relations between them will be coordinative and not oppositional, since paradigmatic ties in this case will prove to be irrelevant, and a simple enumeration of actions on the syntagmatic axis will take place. Syntactic units may also be contrasted to each other at the paradigmatic level. Paradigmatics in syntax has its specific features and cannot be compared with morphological paradigmatics. This problem, however, will not be considered here, due to the purely morphological nature of the problems discussed in this study.[15]

Since grammemes correlating on the paradigmatic axis are a bilateral sign, it is necessary to specify WHAT ASPECT OF THE SIGN IS INVESTIGATED BY THE METHOD OF OPPOSITIONS. As determining the nature of oppositions in a given system or subsystem reveals associative ties between planes of content of grammemes, consequently oppositions serve to reveal THE MEANING of forms and not their outer structure. However, form and meaning are in constant interaction; in some cases the form may become fixed for a certain range of meaning and it can, to a certain extent, reflect more or less explicitly the content it serves to express. Thus, for example, in the word-combination *was reading*, one can find exponents of duration and the past tense, contained as *DF* in the meaning of the given form. But such explicitness of expression is far from being always realized in language, and the form may often be purely arbitrary as, for instance, a case ending. The interaction of form and meaning is revealed in the fact that analogous or similar content will always be expressed in units similar in structure and

[15] Paradigmatics in syntax is discussed in Khlebnikova (1965a) 131-132.

organized in paradigmatic sets. Thus, any series of perfect forms in Germanic languages will consist of homogeneous combinations, independent of the individual invariant of content developed in each separate language.

Thus, correlation of form and meaning is indispensable, since there is no content without form, but the concrete structure of any series of grammemes is irrelevant for the nature of its content and, in the final count, though it is arbitrary, it is not accidental, being always strictly regulated according to the peculiarities of the structure of the given language.

We have established that OPPOSITIONS REVEAL THE PLANE OF CONTENT OF PARADIGMATIC FORMS (grammemes). But it is known that the paradigmatic meaning of morphological units is nothing but the realization of their categorial content, which means that through oppositions we can understand the specific grammatical meaning of the form as a means of expressing some grammatical category.

There exists a voluminous literature on the problem of grammatical category, the review of which would hardly be possible here.[16] Let us adopt the following working definition, proceeding from the premises that a grammatical category expresses relation, is characteristic exclusively of the plane of content of paradigmatic morphological units, and presents a high degree of abstraction from their lexical meaning. Thus A GRAMMATICAL CATEGORY IS A UNIT OF THE PLANE OF CONTENT, REALIZED IN A DEFINITE SET OF FORMS AND REVEALING ITSELF IN THE CONTRAST OF A DEFINITE GRAMMATICAL *DF* (MEANING) AND ITS ABSENCE, OR A NUMBER OF HOMOGENEOUS GRAMMATICAL *DF's* IN GRAMMEMES — MEMBERS OF OPPOSITIONS, JUXTAPOSED ON THE PARADIGMATIC AXIS ON THE BASIS OF IDENTITY OF THE *DF* EXPRESSED.

Taking as initial and given the fact that oppositions are a formal and most objective means of investigating (and a form of realization of) grammatical categories in the morphology of those parts of speech which possess paradigmatics of their own, we can

16 Khlebnikova (1967) 28-33.

give the following definition of opposition in morphology: A MORPHOLOGICAL OPPOSITION IS A SENSE-DISTINGUISHING (OF THE PLANE OF CONTENT) CONTRAST OF TWO MORPHOLOGICAL UNITS POSSESSING A COMMON GRAMMATICAL *DF* (GROUND FOR COMPARISON OR IDENTITY) AND DIFFERING IN THE NATURE OF THE MARKING–NON-MARKING OF THE MAIN INDIVIDUAL *DF*, WHICH DIFFERENTIATES THE MEMBERS OF THE OPPOSITION (BASIS FOR DISTINCTION). THE SENSE-DISTINGUISHING NATURE OF THE MEMBERS OF THE OPPOSITION IS REFLECTED IN THE GRAMMATICAL FORMS OF CONTRASTED UNITS.

It is also customary to ascribe to the sphere of morphology the classification into parts of speech (grammatical classes of words). Are they in opposition to each other? Classes of words are certainly contrasted to each other. They cannot be considered as isolated groups of nouns, verbs, adverbs, etc, unconnected with each other, as is usually presented in 'traditional' grammar.

Parts of speech are linked with each other on the basis of identity and contrast. Thus, the noun is clearly contrasted to the verb by totally formal, syntagmatic, and sense data; synsemantic classes are contrasted to autosemantic, etc.[17] These contrasts, however, are not oppositions, since they have no common ground for comparison. In such cases we deal with contrasting classes of the words of a language, which are naturally heterogeneous and opposed to each other, although capable of intersection. The problem arises here of intersecting–non-intersecting classes which, however, is beyond the boundaries of oppositions.[18] Relations among classes of words are external, generic, and characteristic of all languages. The specific nature of every separate language is made up of inner oppositions on which paradigmatic sets of inflected parts of speech are based. Every such paradigmatic set forms a closed micro-system, the members of which enter regular contrasts among themselves in expressing grammatical *DF*. Every member of an opposition is a carrier of a number of *DF*, which are components of its meaning and are distinguished by their relation to each

[17] On the general problem of parts of speech see Brondal (1943) 28; Trnka (1960) 761-763; Khlebnikova (1963) 36-47; Sunik (1966).

[18] Intersecting classes are discussed in Bogdanov (1965) 3-25.

other. Hence, members of oppositions present clusters of relations in a given microsystem. Oppositions of grammatical *DF* in the system of form-building are symbols of relations without which there can be no system.

Hence, oppositions in morphology may be found only where grammatical categories are expressed, uniting into paradigms the units joined on the basis of some common ground for comparison and differing by the nature of the distinctive feature, which distinguishes members of the opposition.

It remains to be cleared up where the essence of oppositions lies, what types of them may be marked out, and what is the grammatical *DF*, serving as the basis of an opposition.

3.

OPPOSITIONS, THEIR HISTORY AND NATURE

The problem of oppositions and their nature has been worked out in greater detail on the basis of phonology in the well-known work by N. S. Trubetzkoy *Grundzüge der Phonologie*.[1] N. S. Trubetzkoy wrote in a journal on psychology back in 1936 on the applicability of the oppositional theory to grammar,[2] and he used this theory for analysing the Russian morphonological system.[3] Oppositions have since appeared (in open or hidden form) mostly in the works by linguists linked with the Geneva or Prague schools,[4] and gradually, only in recent years, they have been spreading beyond their boundaries. Oppositions have become an inseparable part of the description of phonological units,[5] and were at first used as a method for describing relations within grammatical units only by the representatives of the above schools, the Prague school mainly, which remains loyal to them up to now at any level of linguistic research.[6] As to grammarians of other schools (Soviet ones included), it is only during the last decade that a well expressed

1 Trubetzkoy (1939).
2 Trubetzkoy (1936b) 5-18.
3 Trubetzkoy (1934).
4 For example: Jakobson (1936) 240-288; (1932) 74-84; Karcevskij (1927); Groot (1939) 107-127; Frei (1929); Gougenheim (1939); Cantineau (1952) 11-40; Trost (1939a) 207-211; (1939b) 319-326; Hjelmslev (1939) 51-57. Though Hjelmslev does not belong to the Prague school, he was connected with it at this period, as well as were the French scholars.
5 E.g., Malmberg (1963).
6 Trnka (1961) 129-137 and *Travaux* (1964-1968).

tendency has appeared to make use of the oppositional method in its different aspects.[7]

Besides R. Jakobson and B. Trnka, a most consistent proponent of oppositions in morphology (and binary ones, in particular) is A. V. Isačenko, in whose articles, published in *Voprosy jazykoznanija*, as well as in his book,[8] is considered not only the fact that morphological categories are based on oppositions, but also the nature of these oppositions is analysed. A. V. Isačenko, after R. Jakobson, reduces the types of oppositions in morphology to privative, i.e. the ones in which one element, the weak, unmarked one "does not contain explicit indications as to the presence of the grammatical feature and leaves it unexpressed".[9] A similar approach can be found in the works by R. Jakobson and, in particular, in his article "Signe zéro", which says that the system of noun and verb falls into binary oppositions, one member of which marks the presence of a certain quality, and the other (unmarked) expresses neither the absence, nor the presence of that quality.[10] It is with such understanding of the nature of morphological oppositions that the concept was worked out of the zero sign or feature, of markedness–unmarkedness, of weak and strong members of the opposition, extensive or intensive,[11] and neutral (negative) and positive members.[12]

The rise of such concepts and terms, and their discussion, often contradictory, is due to the clear fact that, in accord with the general development of the science of language, towards the recognition of the need to penetrate into the very depth of the language structure and to study the whole complex of connections between separate levels and their units, it is becoming clearer to

7 Strang (1968); Kahn (1954); Tellier (1962); Ilyish (1965); Mukhin (1964); Khlebnikova (1964a) 150-158. An interesting justification of the application of oppositions can be found in Prieto (1964). In L. Prieto's book many of the ideas put forward in the present edition have found their confirmation.

8 Isačenko (1960) part II. *Morphology*.

9 Isačenko (1961) 36.

10 Jakobson (1939) 145.

11 This term must come from Hjelmslev (1939) 56.

12 Kurylowicz (1949b) 7-15.

linguists that the main type of relations among morphological units are relations of contrast. What is the contrast like; what are its peculiarities in regard to different units; what lies at the basis of the very fact of contrast, and how should it be defined; what are the formal criteria and objective data which would lead to the discovery of oppositions within some units — all these are questions which have not yet found full interpretation and solution in the works dealing with correlations of paradigmatic units.

Language analysts sometimes implicitly proceed from oppositions, without even using this term and suggesting no criteria of their selection and distinction. Thus, for example, in a most interesting monograph by W. Bull, devoted to the category of tense, no special analysis is given of the kinds of oppositions, found in the English or Spanish verb, but the author deems it his main task to reveal the essence of the verb forms in their correlation, and to explain their systemic properties; hence, Bull calls his method systemic as distinct from descriptive, which does not take into account the meaning, or consequently, the systemic qualities of units.[13] The book is imbued with the idea of the contrast of verb sets, considered from the point of view of the markedness or non-markedness of the grammatical meanings distinguished in the system.

Many linguists proceed from the acceptance of polarity and contrast mainly in investigating the verb as a class of words most capacious from the point of view of the complexity of relations among its units. However, the nature of relations among members of oppositions and that basis on which contrast of forms is founded, when turning into opposition, in most cases remain unrevealed.[14] The basis of oppositions — community of the feature or features

[13] Bull (1960) 34.

[14] This can be said about the works cited above, those of B. Strang, A. Tellier, F. Kahn, B. A. Ilyish. The same refers to many other manuals specially dedicated to the description of the verb where oppositions are accepted, more or less explicitly, as a means of expressing grammatical relations among verbal units: Togeby (1953); Hatcher (1951) 254-280; Kurylowicz (1953) 109-118; Diver (1963) 141-181. The same is true of more recently published monographs which will be discussed below.

serving to join contrasting elements into one opposition — is inadequately accented when morphological units are analysed and present themselves as binary signs with expression and content of their own. In regard to phonology things are much clearer, due to the unilateral character of the sign, and not only N. Trubetzkoy, but other phonologists as well (as, e.g. E. Buyssens) strove to underline the fact that community is inevitably revealed in comparison.[15]

And if the authors do point out the nature of an opposition, calling it equipollent or privative, they, nevertheless, do not reveal either the justification of such a division or the nature of the relations of counter-members. Analysing the Russian verb, for example, Van Holk discovers equipollency in Russian aspects without explaining, however, the essence of equipollency and how markedness–non-markedness should be correlated with equipollency.[16] Neither is the reason clear why M. Ruipérez marks the opposition of the Greek present and aorist as privative, as well as the reason for defining the present as a marked member of this opposition.[17] It may be said that detection of oppositions according to relations among their members (privative or equipollent), where it does take place, is of a spontaneous nature and reflects rather the subjective judgment of authors, containing no substantiation of any classification of oppositions.

It should be admitted, however, that most explicitly the theory of oppositions in morphology is reflected in the works by R. Jakobson and A. V. Isačenko, where it is illustrated on actual linguistic material. The maximum clarity is evidently achieved here due to the fact that both linguists introduce the concept of the INVARIANT, without which the theory of oppositions loses its foundation, and they also introduce the notion of *DF*, which lies at the base of any categorial contrast. As is known, in his famous work *Morphological observations in Slavonic declension* R. Jakobson discovers the main *DF* of cases, which correlate in the Russian

[15] Buyssens (1949) 37-38.
[16] Holk (1958) 6-7.
[17] Ruipérez (1953) 241-252.

case system.[18] In his later work, short but very instructive, R. Jakobson constructs a scheme of oppositions in the system of the English verb, where all form types are distributed according to the markedness–non-markedness of certain *DF*. Thus, the form *would have been killed* is marked by the following features: voice, perfectness, and the future tense, and it is unmarked by the feature of the past tense; neither can it have duration and forms with *do*. As R. Jakobson proceeds from the privative nature of all oppositions, he considers the forms of the present and future as unmarked by the *DF* of the past.[19]

The Russian verb is studied as proceeding from privative oppositions in the book by A. V. Isačenko.[20]

Speaking of oppositions as a basis for analysing morphological units, one should mention the book by A. Tellier, devoted to preterite-present verbs and published in 1962.[21] The fruitfulness of taking up these verbs again, the development of which is well enough explained in various manuals of different countries, is ensured in the above book precisely by the investigation of the correlation of modal verbs and their forms on the basis of oppositions, which enable the establishment of inherent laws hidden from the view of a researcher, who would atomistically analyse the meaning and form of every separate verb, outside of their interconnections. This work, however, also considers oppositions as something given; they are simply stated, their nature is not specified, although a wide use is made of the notion of neutralization. In one of the latest foreign works on English grammar, a book by Barbara Strang, correlation of meaning and expression of verb forms is presented as an aspect of binary oppositions of such an order: interrogative–non-interrogative (affirmative) form, non-negative (positive)–negative, active–passive, non-durative–durative, and non-perfective–perfective. These oppositions are based

[18] Jakobson (1958). Cf. the development of R. Jakobson's theory in Revzin (1960) 82-85 and in the books V. Ivanov and V. Toporov (1960); Volockaya and others (1964).

[19] Jakobson (1959) 139-145.

[20] Isačenko (1960) II.

[21] Tellier (1962).

on privative relations of marked–non-marked members. However, they are selected on a mixed syntactic and morphological basis; what grounds they have for comparison is not stated in the book, and a whole number of relations in the verb system as a whole finds itself beyond the oppositions selected, as, for example, relations in the microsystems of non-finite forms and subjunctive mood.[22]

The ascertainment of oppositions without explaining their nature takes place mainly when no problem is raised of the invariant and the *DF* as the basic substrata of any opposition.

The essence of the oppositional method used to consider relations in the morphological system lies in the fact that its fruitful application necessarily stipulates the differentiation of a whole complex of notions denoting events concomitant to oppositions, viz. the notions of zero, markedness–unmarkedness, invariant and *DF*, the notions of neutralization, and the ground for comparison, etc.

Without consideration of all these concomitant notions there can be no basis to declare that any two members are in opposition.

By applying this complex of notions to the analysis, we establish the very points of contact and distinction[23] which characterize the relations between the members of oppositions and, thus, the nature of oppositions is revealed.

This process possesses a logical foundation. The logical side of oppositions is subjected to investigation both by logicians and linguists. Thus, the correlation of the members of oppositions, mainly privative, is studied, from the point of view of formal logic, by A. V. Isačenko.[24] The author shows that relations in privative oppositions correspond to combinations of logically contradictory notions.

Of great interest, from this point of view, are also the attempts to consider oppositions proceedings from concepts provided by

[22] Strang (1968) 161 ff.
[23] On similarity and contrast as a basis of any opposition see for example Martinet (1960); Buyssens (1949) 37 ff.; Slama-Cazacu (1961) 129.
[24] Isačenko (1963) 39-56.

symbolic logic, by establishing the relations between the members of oppositions on the basis of a set theory which helps to discover logical-mathematical relations between the given sets and subsets as members of oppositions. The first work in this respect seems to be the article by J. Cantineau, which was followed by S. Marcus's articles published in Russian editions, dedicated to the problems of structural linguistics.[25] The mathematical aspect of the theory of oppositions, however, will not be considered here, since it is a special question.

The range of discussion of the problem of oppositions in modern scientific literature testifies to the urgency of the problem.

The last decade has revealed an ever greater desire to regulate the methods of analysing the language units, and, hence, to regulate the units themselves. The most fruitful are the works which treat the analysis of the morphological system of any language, taking into account the whole bulk of related problems; i.e. in case the researcher is investigating the *DF* of morphological elements, their paradigmatic invariant, the neutralization of oppositions, etc.

In Soviet anglistics the work introducing the notion of oppositions is the book by B. A. Iljish *The Structure of Modern English*, although it does not contain due consideration of the nature of relations between the members of oppositions, their *DF*, etc.[26]

We may consider as noteworthy the fact that many works on morphology, being published of late, follow the general track, viz., in some open or disguised form they strive to reveal the nature of the contrast of paradigmatic forms, in which is made manifest the content of these forms in their relation to each other in a given morphological microsystem. This happens owing to the fruitfulness of the analysis, proceeding from oppositions as forms of the realization of the meaning of a grammatical category. The method of oppositions is a most objective and formal method of penetrating the content of a paradigmatic system, since the national specific features of the meaning of morphological forms in any language may be revealed in the process of consideration of those

[25] Cantineau (1955) 1-9; Marcus (1962) 22-30; Marcus (1963) 47-74.
[26] Ilyish (1965) 99.

correlations which have been shaped in a given language during its historical development. Hence, it is only natural that the correlations of the meanings of morphological forms are revealed precisely in oppositions, the nature of which should be discovered when studying any system or microsystem of grammatical forms and their meanings.

If, for example, we consider the treatment of categorial contrasting in F. Palmer's work, specially devoted to the English verb, in the section "Uses of the categories" we immediately find the assertion that their use includes both formal and semantic features.[27] The orientation toward observing the interaction of form and content is a most positive quality of Palmer's book, distinguishing him from his predecessors, since many a researcher in the past has shown undue interest toward the formal side and neglected the categorization, or vice versa, an excessive semantization prevailed without considering the inventory of the formal means actually existing in the given language.

F. Palmer draws attention to some of the difficulties and problems (in his formulation) connected with the polysemantic nature of verb forms, when one and the same form, *I'm working*, may be used for denoting a durative action at present (at the moment) and to express a future or habitual actions as in *I'm working tomorrow*, *I'm always working*, (the nature of all those actions is formally expressed in the adverbials). F. R. Palmer finds no way out of this predicament, while by means of oppositional analysis a scientific explanation may be found of the potential capability of verb forms to have seemingly diametrically opposite meanings. Such an explanation may be obtained through formalizing grammatical semantics, which is achieved, as will be shown further, by the oppositional method, through establishing the invariant of categorial forms and their *DF* on the basis of their systemic correlations.

As a direct result of the absence of the notion of the invariant in his work, F. R. Palmer does not include the future with the

[27] Palmer (1968) 59-60.

verbs *shall* and *will* in temporal correlations, a tendency which started with O. Jespersen and is supported by some linguists today without any due justification.

Besides the lack of any clearly defined invariants of verb forms, the reason for neglecting language events and excluding the future from the system of categorial forms lies in two more factors: (1) lack of a criterion of the analytical form as a morphological and not syntactic fact, (2) a tendency, stemming from descriptive linguistics, to consider the forms from the point of view of their linear structure and not their categorial content. Let us explain these two factors: strange though it is, the problem of the analytical form is best developed not in foreign linguistics, based on languages abounding in analytical forms, but in the Soviet language science.[28] One of the decisive criteria of the analytical form, as an equivalent of the synthetic one, and member of the paradigmatic set, has always been considered the desemantization of the first component and the idiomatic nature of the whole combination. However, the determination of the degree of desemantization and idiomatics to a great extent depends upon the researcher's intuition, and here the necessity arises for more formalized criteria, no matter how difficult it may be to formalize the meaning of grammatical forms.

We may suggest the following way of discovering analytical forms: (a) an analytical form is present where it correlates, in an opposition (the basis for which is identical *DF* of a grammatical category) with another form, correlating with it in the same categorial plane, and where it is distinguished for its universality and unlimited reproduction within the given category. Thus, the

[28] The peculiar properties of an analytical form and its auxiliary part have been observed as far back as the beginning of the century in the Moscow school of linguistics: Fortunatov (1956) I, 178 ff. The analytical form is discussed in the following articles of Russian linguists: Jarceva (1953); Smirnickij (1956) 41-52; Zirmunskij (1961) 3-21; Khlebnikova (1966b) 44-50. In linguistics outside Russia the problem has not been as widely treated as could be expected, due to the analytical structure of the languages described. Analytical forms are mentioned in Zawadowski (1959) 112 ff.; Tesniére (1939) 153-183. Strang (1968) 154 ff. includes what we call analytical form under the item of complex conjugation.

forms of the future tense correlate, by the *DF* of the grammatical category of time (tense), with the forms of the present (or past), entering parallel paradigmatic sets: *I do — I shall do*; *I slept much yesterday — I shall sleep much tomorrow*; (b) the desemantization of the first component of an analytical form is correctly determined (with the first criterion observed), if there exists a definite set of contexts in which the general meaning of the utterance is incompatible with the full meaning of the first component. In the sentences, for example,

> *You will come to me, won't you.*
> *They will do as they are told.*
> *I told you I shall do as you wish.*
> *I shall do it willingly, etc.*

it is difficult to find any tangible retaining by the auxiliary verbs of the initial modal meaning of wish and duty. Besides, because the verbs *shall* and *will* have joined the system of conjugation and lost their modal content, they no longer belong to the system of modal verbs, and we may speak only of separate cases of modal use of these verbs, the main purpose of which in Modern English has become purely grammatical. *Shall* and *will* no longer enter temporal correlations within the microsystem of modal verbs (the modal verb *should*, for example, does not correlate with *shall*, and *would* in its modal use with *will*). They have become exponents of tense in the system of categorial relations of the synthetic-analytical members of the paradigmatic set of the English verb.

The same situation is found among the verbs of similar function in other Germanic languages.

It stands to reason that the isolation of analytical forms of the future from the temporal paradigm, viz., joining various verb combinations with different categorial meanings exclusively according to the linear pattern which they present (e.g. the joining of *would have* with *can have*, *might have*, etc.), stems mainly from the insufficiently developed criteria of the analytical form, mentioned above, and, naturally, from the imperfection of the oppositional analysis applied by different authors.

Thus, F. R. Palmer refers all the verbs, which are modal by origin, meaning and structure to the secondary (modal) auxiliary verbs,[29] evidently, by the types of syntagmatic pattern which they form. The patterns *could come, was to come, will (would) come* have indeed homogeneous components and the same linear structure. It is clear, however, that their functional load is different. Some combinations, such as *could have come*, should be considered at the syntactic level, since they consist of two notional components; others should be considered at the morphological level as members of paradigmatic sets — means of expressing abstract grammatical categories *(will, would come)*. The lack of such differentiation may lead to the mixing up of the units of different levels of language structure and to complete chaos in defining the morphological inventory of the verb system. So, the English verb in this case will lose all the forms expressing future indefinite, conditional, and subjunctive, i.e. the sets which are quite important communicative units of the language.

A review of only one question linked with the interpretation of verb categories shows how important it is to define a system of notions and rules, which regulate the structural organization of any side of the language. If such rules are not defined, the interpretation of grammatical forms will be arbitrary. An example of this may be found in the scheme of categorial relations of the English verb, postulated by F. R. Palmer, who thinks that the tense has three functions: (1) to denote purely temporal relations of present-past, (2) to carry out the function of the sequence of tenses, (3) to express unreality in conditional sentences and in wishes.

It may be said about the scheme that the function of the sequence of tenses is also carried out by forms of the type *should (would) do*. and F. R. Palmer does not include them in any paradigm; such a system completely suspends the question of the means of expressing unreal moods and future tense. The latter, however, is clearly present in English and is well enough differentiated, expressing a future action which is considered from the positions

[29] Palmer (1968) 15.

of present and past. This is well confirmed by the most elementary language data, and to withdraw from the theory of language that which exists in the language itself is hardly expedient; it would be more logical to try and find a solution to the controversial questions, and to define a place in the system of units found in the language. An overall oppositional analysis which calls for a homogeneous determination of a number of basic and secondary language events would be of great help for this purpose.

F. R. Palmer's scheme consists of two units — present and past, which intersect binary points of perfect–non-perfect and progressive–non-progressive.[30] Thus is given a binary cross-section of the verb paradigm (which is becoming the norm for most authors trying to analyse the verb), but the nature of oppositions and correlation of their members is not investigated, for lack of a preliminary determination of those notions mentioned above connected with oppositions, which should constitute an organic part of a morphological analysis.

The book by M. Joos, published a year before, was a step forward compared with F. R. Palmer's work. M. Joos devoted a quarter of a century to the description of the English verb, which finally led him to oppositions.[31] He not only does binary cross-section of the verb formtypes, but also tries, for the first time, as far as we know, to establish the nature of oppositions. Thus, for example, M. Joos considers the progressive, as well as passive, form to be a privative member, and introduces the notion of a marked–unmarked member.[32] However, the determination of the privative nature of some members is to a certain extent intuitive; it is not verified by neutralization, neither is a question posed on the invariant of each set and its categorial *DF*.

Besides, M. Joos' notion of a privative member does not cor-

30 Palmer (1968) 59.

31 Joos (1964) 112.

32 The fact that some of the notions adopted in the present edition appear in M. Joos' book is an indication of the fruitfulness of the oppositional analysis and of a certain community of ideas among contemporary linguists. Another confirmation of this is the book by Allen (1966) 24. This book, however, does not contain any oppositional analysis of the kind offered here.

respond to the logical interpretation of such a member as devoid of the *DF* present in the counter-member of the opposition.[33] Hence, all these problems await their interpretation.

[33] Cf. Ogden (1932) 47.

4.

THE INVARIANT AND DISTINCTIVE FEATURES (DF)

Since at no matter what angle the oppositions in morphology are considered we have to deal with the notion of invariant and with the basis of this notion — distinctive features, the need arises to dwell in detail on these questions, and all the more so, that up to now there exists no clear idea on their importance for morphology.

Every grammatical form possesses, a basic, i.e. general systemic meaning — the invariant,[1] which is the core of relations with other forms and upon which oppositions are based. On the other hand, each form may acquire in speech some combinatory variants of the meaning, caused by the distribution of the form and of its lexical content. The invariant is the centre of the cross-section of relations in a microsystem composed of oppositions based on grammatical *DF*; it is a quality of the language as a more or less static inventory of signs. On the syntagmatic axis both invariant and variant meanings are realized, but in any use of a variant meaning there is always present the invariant of the grammatical

[1] On the relation of the basic *(Grundbedeutung)* and general, invariant *(Gesamtbedeutung)* meaning, and on the origin of the problem see Jakobson (1936) 240-44. The problem of the invariant is dealt with in Leška (1964) 85-93.

Bazell (1966) 333 expresses the following view about the fruitfulness of the notion of the invariant for the verb: "Tense and aspect are the verbal inflections with maximal intercommutability of morphemes .. Here, if anywhere, is the place to look for Gesamtbedeutungen with a reasonable chance of success."

content of the form. The notion of the invariant is important, since it helps establish the nature of oppositions within the paradigm, and select those contrasts of grammatical signs, regularly repeated in parallel sets, which are relevant for the system and form its basis. The invariant indicates only the relevant distinctions in the language system and helps the analyst to abstract himself from the lexico-stylistical and other contextual peculiarities of what we call usage. But the invariant does not at all presuppose a monosemantic meaning of the grammatical form; the fact is that every morphological unit includes not just one component of meaning, but a number of components (semes), which are its *DF*, just as, for example, the features of nasality, labiality, and sonority are combined in one phoneme.

Thus, every verb form, which we call a grammeme, is a carrier of grammatical meaning consisting of distinctive features — semes. It is the *DF* relevant for the given opposition which enter regular relations in the paradigm. For example, for the opposition present perfect-present perfect continuous of the English verb *(I have done — I have been doing)*, relevant will be the *DF* (1) of the axis of orientation — present, (2) of anteriority to the axis of orientation — perfect, (3) of duration, *DF* always imply correlations on which any morphological system is built.

The invariant is defined depending on the place the given grammeme occupies in the opposition within a grammatical category.

If, for example, we know from all school grammars that the form of the present indefinite expresses an action taking place at the moment of speech, a past action, a habitual action, a future one, etc. then, on the basis of such data on the content of the form we shall not be able to judge its systemic, invariant meaning, which ensures its definite place in verb oppositions. Of all the variants, we can extract the invariant not mechanically but through some definite thinking operations, taking into consideration the possible nature of the oppositions into which enters the given form in the given language. The variants of the meaning of a grammeme are always combinatory and depend either on syntagmatic data or on the lexical content of the form; they may also have a stylistic basis.

Concerning the notion of variation in morphology, it is necessary to distinguish the different meanings of the term 'variant' as applied to verbal grammemes: (a) morphological units possessing one and the same grammatical meaning may have different forms of expression (cf. various types of declensions and conjugations in highly inflected languages). In such cases morphological variants concern the plane of expression, the form of the unit; (b) in connection with the notion of the invariant, variants denote variation in the meaning (plane of content) of the grammeme.

The invariant is a unit of the plane of content on THE PARADIGMATIC AXIS, in the system, whereas any variation in the general meaning of a grammeme occurs on THE SYNTAGMATIC AXIS, under the influence of the environment.

It is because modern linguistics shows a tendency to comprehend the qualities of the inner structure and the deep-seated relations of the object under study that, in describing the functions of units of different levels of the language, researchers are increasingly resorting to the problem of the minutest indivisible *DF* of different language units.

It is common knowledge that the most fruitful proved to be the discovery of similar features on the phonemic level, and it is on these features that the modern theory of the phoneme is mainly built, the phoneme being but a cluster of contrasted distinctive elements, correlating with each other in different ways.[2]

If the phoneme is a cluster of such elements, then the very fact of their existence implies correlations based on the opposition of either the presence and absence of certain *DF* or of different degrees of the quality marked by *DF*, the correlation of *DF* always implying their hierarchy, since equivalent relations find no reflection in oppositions and in the paradigmatic system generally. This rule stems from the fact that any system (or microsystem) of signs,

[2] Cf. the works of Jakobson and his co-authors, and also Reformatskij (1961); Kuznecov (1958) 55-61; Malmberg (1963) ch. VI; Šaumjan (1962) 121 ff., 130 ff.

being a paradigmatic extraction of linear chains, is economical and excludes the coexistence of units (or their minutest features) with an identical communicative or sense-distinguishing function in the system itself.[3]

The detection of *DF* in the phoneme is facilitated by the unilateral nature of that sign,[4] which has expression but has no content in the plane of correlation with categories of thought. Correlations of *DF* are, so to speak, the 'content' of the phoneme, its quality. Such phonological *DF* as 'voiced–voiceless', 'open–closed', 'long–short' have no special exponents in the plane of expression.

At other levels of language structure, the units of which correlate with categories of thought, the detection of the minutest, indivisible elements is connected with certain difficulties, since a problem immediately arises of the correlation of the form and content, thus giving rise to conflicting interpretations. However, penetrating deep into the inner structure of bilateral signs is essential for determining their systemic qualities for each language, and here the need arises to look for ways of formalizing the linguistic analysis. One of these ways is decomposition of the unit into its components (componential analysis on the grammatical level), and the consideration of its place in the system on the basis of correlations with other similar units.

In discovering the minutest elements in morphological units much can be gained, (with respect to the methods of analysis) by the application to morphological material of the results obtained in the course of phonological research, for no matter how the problem of isomorphism is treated,[5] one cannot deny a certain parallelism in the organization of units at different levels of linguistic structure. That is the reason for successful application of homogeneous methods of research (such as distribution, substitution, opposition, transformation, etc.) to different objects, be it

[3] The idea of economy in language is developed by Martinet (1960) 182-84.

[4] 'The second articulation of language' according to A. Martinet (1962) 26.

[5] Kurylowicz (1949a) 16-26.

lexicology and semantics, the phoneme, morpheme, word-form, or syntaxeme.[6]

Considering the question of the minutest components of a categorial form, one should first of all turn to the statements by V. Skalička, the first linguist to have posed this problem and to have introduced the term 'seme' for such events. In his *Hungarian grammar*, V. Skalička states that if grammar is a system, then we should look for the minutest units constituting the system.[7] Because morphological units possess some meaning, their minutest *DF* received the name of 'seme'. To avoid mixing up the categories of lexical and grammatical semantics (a seme is a component of the 'semanteme' or 'sememe')[8] it is convenient to mark the components of the meaning of a morphological form by the term 'morphoseme', suggested by O. S. Akhmanova. The term 'seme', however, is also possible since we are dealing with elements of the plane of content.

Hence, the morphoseme is the minutest, no further indivisible component of the meaning (content) of a morphological form (or: categorial form, grammeme).

The singling out of morphosemes in each member of a paradigmatic set helps avoid a subjective definition of the content of this or that word-form, a tendency which still dominates grammar and often leads to inconsistency in understanding the function of some paradigmatic set in the system of a language. The determination of morphosemes is a means of formalizing grammatical semantics, a field more difficult to formalize than the unilateral phonemic unit.

One may object to this means, for if the morphoseme is a component of the plane of content of a grammeme, and not of its formal composition, what guarantee can we have that in this case as well, the discovery of the minutest signs will not be per-

[6] The notion and term 'syntaxeme' are used by: Mukhin (1964) 3 ff.; Šabršula (1966) 188.

[7] Skalička (1935) 12. R. Jakobson poses the problem of morphological features in one of his earliest articles Jakobson (1936) 240.

[8] Akhmanova (1961) 117.

meated with the subjective judgment of the researcher? A truly formal analysis of the plane of content of a morphological form may be ensured only on condition that all morphosemes are selected on the basis of those correlations, on which any morphological system is founded, that is, when they are marked out in terms of the relations and contrasts which the members of any paradigmatic set are bound to enter. The detection of morphosemes, common for the correlating pair and distinguishing the members of the given pair, is that very formal factor to be relied upon when dealing with the semantic aspect of a morphological form.

If, for example, we analyse a case form of the noun or some tense-aspect grammeme of the verb, the components of their grammatical meaning may be defined on the basis of those contrasts (oppositions) within these grammatical categories to which the given form is liable as a member of a definite paradigmatic set in the given language. Thus, the first principle we have to consider when analyzing morphological units is the necessity of their entering the oppositions of contrasted grammatical *DF*. For instance, the grammatical content of the Russian word-form *stolov* is revealed only through correlations of the *DF* of number, gender, case, with the grammatical *DF* (morphosemes) always intersecting in one form. The categorial meaning of a grammeme may be defined only as a result of the study of the whole totality of *DF* pertaining to the given part of speech in the given language.

The word-form *stolov* contains three morphosemes, since this part of speech in Russian expresses three grammatical categories — number, gender, and case. One should always be aware of the fact that every morpheme is able to express more than one grammatical *DF* in modern languages (shown by V. Skalička). In most cases a single form is a cluster of *DF*, as in the above example of a Russian case-form. That is why the morphemic analysis is inadequate for objective investigation of the essence and properties of a morphological unit.

Another example: the suffix in the Russian verb-form *piš/eš (thou writest)* expresses the present tense, imperfective aspect, person, number (marked distinctive features) and a number of unmarked *DF*, such as active and indicative, revealed at the level of interparadigmatic and not intraparadigmatic correlations. Marked morphosemes *(DF)* constitute the inventory of grammatical categories expressed in the paradigmatic set of an inflected part of speech. They always have formal exponents, but it does not, of course, mean that every single seme has its own exponent. Far from it. In most cases inflectional formatives (both synthetic and analytical) stand for more than one *DF* (as in the given examples), and, accordingly, within the plane of content, every grammeme (or formative) represents a cluster of categorial *DF*.

Hence, if we have no 1:1 correspondence between the morpheme and the components of its content, then the only way of objectively analyzing the plane of content of grammemes will be through the analysis of categorial contrasts, entered by the given grammeme in a given language, and on the basis of which *DF* (morphosemes), the initial and final points of oppositions,[9] may be detected.

If we admit the morphoseme as the minutest unit of the plane of content of the grammeme, we should specify which content is meant — paradigmatic or syntagmatic. In the mixing up of these axes of the functioning of a grammatical unit are hidden numerous errors in the determination of the content of morphological forms, while a strict demarcation of these functions (associative and linear) are a guarantee of the highest objectivity in solving the problems of the plane of content of grammatical units.

The determination of morphosemes and the invariant of the form as a member of a paradigmatic set, on the one hand, and the investigation of the nature of the realization of this invariant in speech, on the other, are different things calling for different methods and producing different results. Thus, for example, what idea can we get of the place in the paradigmatic system of the English present perfect and its invariant if we say that it includes the morphoseme of the future (in the example *after I have finished eating, I shall start dressing*) or durative aspect (in *I have worked*

[9] Cf. Spang-Hanssen (1959) 19.

here many years); what is the use to assert, in building paradigmatic invariants, that the English present indefinite has semes of the future *(I leave tomorrow)*, of the past *(suddenly I see a man rushing out)*, or of durative aspect *(see, how I take the glass, pour some water into it ...)*.

It is neglecting the differentiation of the paradigmatic invariant and of syntagmatically or lexically conditioned variants that creates a confusion in grammar and hinders the study of the systemic ties between the units of grammatical microsystems.

It is natural that in a mixed syntagmatic-paradigmatic approach there can be no question of an invariant of a paradigmatic set. Such a confusion of associative and linear functions of morphological units happens rather often in linguistics. As a sample of linguistic description based on differentiation of paradigmatic and syntagmatic axes, we may cite two works by Russian linguists: V. V. Ivanov and V. N. Toporov, *Sanskrit*,[10] and Z. M. Volockaya and others, *Opyt opisanija russkogo jazyka v jego pis'mennoj forme*,[11] where the notion of *DF* stems from the idea of the invariant of a paradigmatic set.

As the authors of the latter work rightly point out, by means of such *DF* there can be compiled an economic and convenient scheme of description of the category of case. The same equally concerns paradigmatic schemes of any part of speech, and the more complicated the scheme (as, for example, the verb paradigm saturated with heterogeneous grammatical categories) the more useful will be the componential analysis of the grammatical meaning into *DF*. Such a procedure meets the requirements of formalization. Such formalization on the level of grammatical semantics, however, proves effective if the morphosemes are determined exclusively on the basis of correlations, i.e. if every seme in the word-form is linked with a contrasting seme in another word-form, compared or contrasted on the basis of an identical grammatical category (on the paradigmatic level). Each seme is necessarily the initial and final point of the opposition based on contrasting

10 Ivanov and Toporov (1960) 77 ff.
11 Volockaja and others (1964) 36-38.

grammatical features. It is the intersection of *DF* in any unit of a paradigmatic set and the primacy of correlations as a formal mark of the nature of associative ties in a microsystem that are insufficiently emphasized even in the works which do raise the problem of morphological *DF*. V. Skalička, for example, does not raise such a problem in his above-mentioned work.

Opposition is characteristic exclusively of the paradigmatics of signs, which is an extraction from a system of regularly organized homogeneous units, joined by the identity of their generalized grammatical content. Paradigmatic units, despite their belonging to the units of one order, are always heterogeneous and cannot have an identical communicative content; the act of speech, then, is a two-choice situation between elements of the paradigmatic system, according to the purpose of the utterance. That is why morphological paradigmatics lacks synonymy of forms with identical categorial content (and a set of *DF*). although neutralization may take place of one of the members of the opposition — rapprochement on the basis of contrasting. Neither are there any phonemes with an equal set of *DF*, though rapprochement on the basis of neutralization may take place as well, under certain syntagmatic conditions.[12]

The nature of correlations in the paradigmatics of one or another level of the language structure depends on the properties of the units of a given level, and there can be no complete isomorphism in this case. In other words, the nature of relations among the units of a phonemic set is different from that of the word-forms, and paradigmatics in syntax has different specific features due to the primacy of linear relations in syntactic units (members of the sentence, word-combinations or syntagmas, and sentences). The interrelation of paradigmatics and syntagmatics also changes according to the nature of the unit.

Since in this case we are interested in the principle of the analysis of the plane of content of morphological units as members of a

[12] The differentation of synonymy, variation, neutralization in morphology, and similar facts related to them is discussed in Khlebnikova (1970) 3-23,

paradigmatic set, we shall confine ourselves to the questions related to this task.

In the units of a morphological system, contrasted on the basis of identity of their general grammatical meaning, it is morphosemes that correlate among themselves. Does this mean that one grammeme may enter a correlation by one grammatical feature only, i.e. that it enters one opposition? Certainly not. One grammeme always includes at least two morphosemes, and in the verb, the most capacious class of words, a considerably greater number of morphosemes may be contained in one grammeme, depending on what grammatical categories are expressed by the given part of speech in the given language, the oppositions stemming from each morphoseme separately. The formal criterion for singling out semes, as was shown above, is a correlation of the seme detected with another seme or with its absence in the contrasted grammeme, provided there is a general ground for comparing the members of the opposition. The greater the number of grammatical categories expressed by the given part of speech in the given language, the greater the number of semes which are contained within each member of a paradigmatic set.

Morphosemes, making up the plane of content of a grammatical set, form a hierarchy and are not equivalent; accordingly, different semes are to a different extent relevant to oppositions depending on the value of the grammatical category they represent. Thus, for example, within the verb set (by which we shall mean a set of verb conjugation[13] in all the persons of one tense-aspect, mood or voice), the semes of person and number, constituting an indivisible, general grammatical category of person-number will not be relevant for correlation with other sets of the verb system, but rather the semes of the more abstract categories of tense, aspect, mood, and voice will be relevant. It is not without reason that person-number are called "concord categories"[14] or "local accident".[15] Semes of person-number are not basic, but attendant ones. It is

13 Cf. the use of the term 'form-set' in: Strang (1968) 143.
14 Strang (1968) 143.
15 Martinet (1962) 52.

evidently due to this factor that the formal markers of this category are easily reduced in the process of historical development of the language, given some real prerequisites in its history. Thus, for example, in English, in spite of the fact that it has a most differentiated tense-aspect system, the forms of concord in person and number have been nearly reduced to nil.

The question of the degree of relevance-irrelevance of different morphosemes for oppositions should be solved separately in relation to every language. The interaction of strong and weak semes in different oppositions, the nature of their interdependence, and their compatibility — incompatibility are all problems to be settled separately for each language.

Thus, the Russian perfective aspect has no seme of non-duration since duration is not a paradigmatic *DF* of Russian verb-forms, as it has no special mark. The imperfective aspect will have no seme of non-perfect since perfect, as a marked *DF*, is lacking in the Russian verb. In English, on the contrary, in the indefinite set *I write* the weak seme of non-duration is present, since this grammeme is contrasted to the grammeme *I am writing* by the *DF* of duration, which is marked in the plane of expression and content. The presence of such a seme in the unmarked set is fully confirmed on the syntagmatic axis in the process of neutralization of oppositions, which will be discussed in corresponding sections of this work.

Strong semes are always marked in the plane of expression. This does not mean, however, that markedness cannot be expressed in a zero formal marker, but the lack of a formal exponent can be understood only from the correlation with other members of that very paradigmatic set where the same semes will find expression in a special morpheme (cf. Russian *rek* where the genitive case plural has a zero exponent and *stol/ov*, where the same case is expressed by the suffix *ov*). Hence, the interdependence of the units of the plane of expression is also defined on the basis of correlations.

A totality of semes makes up the invariant of a paradigmatic set or word-form, the interaction of morphosemes leading to a constant intersecting of oppositions in each member of the paradigm.

This may be illustrated by the intersecting of semes in the English past perfect continuous (for details see chapter 14).

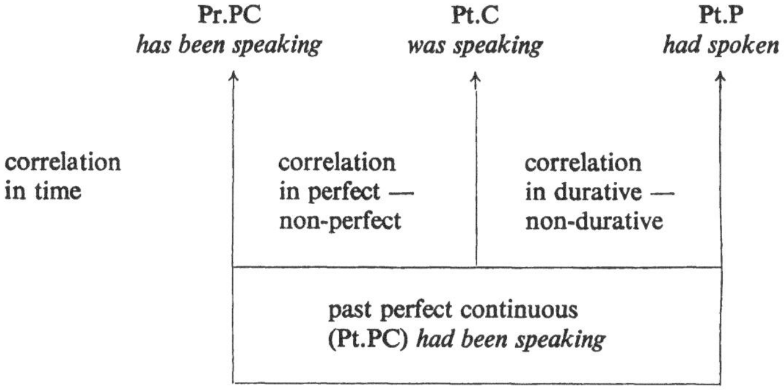

The above correlation preserves its basis for comparison in the aspect of two common strong semes — time (tense) and perfectness. Not every set, however, gives similar correlations. The correlation of strong and weak semes depends upon the specific nature of the grammatical content of each set.

It follows from the example of componential analysis of the English verb form that the morphosemantic analysis is valid for any language. It is also clear that person-number plays no part in the above correlations; the correlation of forms of different person-number is contained within one set (such as the present indefinite, for instance) and does not go outside it, being irrelevant for the essence of oppositions existing in the paradigmatic set of the given verb system.

Morphosemes are differently combined not only in different verb sets, but in various microsystems as well (of the voice, subjunctive mood, and verbals).

Studying concrete correlations in different microsystems reveals which morphosemes are strong (marked), and which ones are weak (unmarked), depending on the nature of the oppositions entered by the morphological sets.

5.

THE ZERO

In connection with the above we should clear up the question of the zero member of opposition, and of its role in a morphological system. We can agree with E. A. Makaev[1] that the introduction of the concept of a zero morpheme into linguistics (as well as the introduction of neutralization), has opened the way to structural research, and we may add, to the analysis of oppositions. We should specify, however, (and that is essential) the notion of zero in the plane of expression and in the plane of content. These two different things should be differentiated, something which is not always done.

As was pointed out above, the zero in the plane of expression is defined by the place of the word-form in the system of grammatical inflections. Thus, in the Russian word-form *rek*, the zero inflection is singled out in comparison with other forms of the same root, of the type *rek/a* (nominative sing), *rek/oj* (ablative), *rek/i* (genitive), etc. Comparing it with all the word-forms of the given word as a part of a certain class of declension, we define what grammatical meaning is expressed by the zero morpheme. In this way we find out that the given member of the case correlation is defined by the absence of a special formant, i.e. by zero. The concept of the zero morpheme or zero morph as a unit, contrasted to some other unit in the plane of expression, has long

1 Makaev the answer to the questionnaire offered by A. Martinet, *Travaux* (1957) 69-70.

since come into use in linguistic research, and American descriptivists also use it.[2]

The zero problem is elucidated in greater detail in G. Meyer's book *Das Zéro-Problem in der Linguistik*. The book gives the whole history of the problem and raises a number of important questions related to the zero problem, for example, whether there can be the zero of function, as well as of form, and whether it can be studied paradigmatically, or syntagmatically as well. Meyer denies the need to analyse oppositions in a system, though this opinion is not well enough supported by facts, and he often resorts to oppositions himself. From this conception follows the absence of the zero differentiation in the plane of expression and the plane of content, which leads G. Meyer to deny the relevance of the zero concept in the formal analysis of the system.[3]

A rational solution, in our opinion, is offered by B. Pottier, who thinks that zero is present where the formant is not expressed, entering the class of categorial markers *(catégorisateurs)* and concord formatives.[4]

The article by R. Jakobson, devoted to the zero sign, does not differentiate the zero as an absence of the formative to mark some category and the zero category as a member of an opposition in which the contrasted feature is not marked. Thus, R. Jakobson says that a grammatical system acts on the basis of opposing something to nothing ("avec rien").[5] From there, according to R. Jakobson, results the zero aspect (imperfective), the zero tense (present), and the zero number (singular), i.e., zero is considered to be a member in which the contrasted feature is not marked. But it is one thing that the singular is expressed through the absence of a formal marker, and another that it is unmarked in its content as a member of the opposition in number. In forms of the singular

[2] The zero exponent has been discussed by many authors of different linguistic persuasions: Peškovskij (1956) 16; Bloch and Trager (1942) 59, number the zero expression among morphological processes. On zero see also: Frei (1950) 161-191; Harris (1951) 181; Godel (1953) 33-41; Haas (1957) 33-53.

[3] G. Meier (1961) 89 ff., 279 ff.

[4] Pottier (1962) 165.

[5] Jakobson (1939) 144.

the grammatical number is marked by content, but not by form, i.e. it is characterized by the absence of a marker (zero of expression). In the same way it is the present tense that is always expressed in the form of the present and not any other tense, and the forms of the Russian imperfective aspect express precisely this aspect and no other. These categories are formally marked by zero. Hence, we can agree with A. V. Isačenko that the meaningful absence relates to the designator and not to the event which it designates, and that there can be no zero category.[6]

A. M. Peškovskij also discusses the zero category in relation to the positive degree of comparison.[7]

It is advisable, however, to distinguish the following: (1) when a special formal marker of grammatical meaning is absent, its absence is distinguished as compared with the inflections of other forms of the same word or word class, which enter one common set of forms of the grammatical category. In such cases it is the zero morpheme or morph, as in the genitive case of *rek*, and in the singular in nouns in which the singularity has no formal expression. It is in such cases that we can speak of the zero or, rather, of the zero in the plane of EXPRESSION — the sphere in which the zero concept is valid; (2) when one member of the opposition does not express within its CONTENT the categorial feature contained within the second member of the opposition.

In this case we can speak of the unmarkedness of this member and not of a zero category, meaning by 'unmarkedness' precisely the plane of content and not of expression, as, for example, is the case in the oppositions of indefinite and continuous sets in English, where duration is unmarked in the indefinite set and is absent as a semantic component of this set. Here we deal with unmarkedness and not with zero, since in this opposition the *DF* of duration is relevant, being marked formally and in content in the strong member of the opposition; that is why the weak member (the indefinite set) can contain no formal expression of this *DF*, and we can only state the unmarkedness of the given meaning; (3) when

7 Isačenko (1960) 32.
6 Peškovskij (1956) 29.

the unmarkedness of a member of an opposition is combined with a zero marker, i.e. with the absence of a formal designator of a category. Then the unmarkedness has a double nature — it means both the absence of a formal sign (zero of expression) and the absence of a corresponding categorial meaning.

For example, let us take the common case in English and the zero modality in the indicative mood. Here, the grammatical category on which the opposition is based (case, mood) is not 'formally' marked at all, and 'in content' the *DF* of case or mood (modality) is also unmarked in the weak member of the opposition.

So, we shall call a zero the state when a *DF* of a grammatical category is not expressed formally (has no formal exponent), though the *DF* itself belongs to a unit — member of a paradigmatic set of the given class of morphological elements. Zero is an implicit sign, understood through its relation to other units of the given paradigmatic class.

6.

THE NATURE OF OPPOSITIONS IN MORPHOLOGY

Since oppositions reflect the relations of the plane of content of morphological units (grammemes) having a certain formal expression, it stands to reason to try to find out the nature of oppositions possible in a morphological system.

From our point of view not all oppositions in morphology can be reduced to privative. By privative oppositions we understand (according to the logical definition) a relation in which a privative member implies the absence of the feature expressed in the counter-member of the opposition.[1] The privative member, then, is an unmarked, weak member of an opposition. What oppositions will meet such a definition? First of all INTERPARADIGMATIC ones. It is natural that different paradigms of the same morphological class cannot carry one and the same *DF*. That is why in an interparadigmatic opposition one paradigm expresses this *DF* and the other is contrasted to it precisely by the absence of the given *DF*. Thus, for example, in the field of voice, one set of forms shows passive voice and the other does not. This voice is not marked in the active, either in form or in content. Hence, the active is distinguished by way of contrast to the passive, as an unmarked member related to a marked one. In the same way, the subjunctive mood is contrasted to the indicative as a marked member to an unmarked one. The members of these two interparadigmatic oppositions are

[1] Ogden (1932) 47.

by themselves multinuclear, and each one in its turn consists of a number of oppositions. Hence, in the line of the grammatical categories of voice and mood, not isolated forms or homogeneous sets of forms are contrasted, but multinuclear units, which are closed sets of forms also linked among themselves by definite relations. Within one paradigmatic microsystem there may also exist a privative opposition, though not in all cases. Thus, if we compare the forms of the indefinite set *(I write)* with a continuous one *(I am writing)*, it becomes clear that this is a purely privative opposition, since duration, marked both in form and content in the continuous set, is not marked in the indefinite, which, since its contrasted *DF* is not marked, is a weak member of the given opposition. On what basis, then, are these two members joined ino one opposition in aspect (durative)? The basis for comparison here (or common *DF* for both members) is the general temporal plane of action, both in the plane of the axis of orientation (present) and in the plane of relation to the given orientation axis — simultaneity with it. Through the community of the temporal plane in the sphere of anteriority (priority to the time of the axis of orientation), oppositions of the perfect and perfect continuous sets *(I had written–I had been writing)* naturally belong to such privative oppositions.

The fact that these formtypes are joined into pairs of privative oppositions is confirmed by the isomorphism in their content, which leads to some common points in their usage, first of all in the process of neutralization, to be discussed below.

Not all oppositions, however, are privative through the logical relation between their comembers. In this connection, we cannot agree with the interpretation of present tense as a form in which the real temporal relation of the action is not expressed, as shown in A. V. Isačenko's article "On grammatical meaning",[2] and we cannot consider the form of the present as an unmarked member of the opposition present—past. Both present and past express relevant *DF* in the plane of content of the grammatical category

2 Isačenko (1961) 34.

of time (tense), and both present and past are correspondingly marked by content, as well as by form. *I look* compared with *I looked* has a zero exponent, while in Russian *smotr/ju–smotr/jel* the forms are marked in both the planes of expression and content.

The action of a verb in the present always coincides with the moment of speech, but it can take place within the range of a different period, which, however, always includes the concept of 'now'. Even in the sentence *the earth rotates*, where the present tense does not imply the limits of the duration of the action, the moment of speech is clearly included into the action. That is why, considering the present as a weak member of a privative opposition, as a non-past (in the same way as a non-durative action is contrasted to a durative one in the privative opposition of non-continuous–continuous forms) would simplify the essence of temporal contrasts. Members of such contrasts are logically equal, both expressing *DF* relevant for the given opposition. However, in every such opposition one member can be more general (weaker), as the initial member of the opposition (in this case the present), and the other more clearly marked (a strong one), as a counter-member of the opposition (the past). Absolutely equivalent members of an opposition hardly exist, since in any opposition a certain hierarchy of the members is natural. Of course, we can conceive a purely abstract idea, that present is not past and, thus, recognize the privative nature of the opposition, but this way will not bring about the understanding of the nature of the distinction in the relations within the given contrast, and, what is most important, the syntagmatic actualization of such an opposition in the context and the relation of its members to neutralization testify to the inconsistency of such an interpretation.

Temporal oppositions may be considered as equipollent (according to N. S. Trubetzkoy's classification), since both members of such an opposition are characterized by their own specific quality.

The trend to discover the varied nature of morphological oppositions; not reduced to privative, manifests itself in some works on grammar. Thus, B. Trnka in his article devoted to the principles of morphological analysis calls 'equipollent' those oppositions in which every member is distinguished by its own morphological qualities, as, for example, the contrast of nominative and vocative cases in Greek and Latin or of feminine and neuter gender.[3] J. Cantineau calls 'equipollent' an opposition of masculine and feminine gender, as well as the contrast of *goose* — *geese* by the purely formal sign of the presence of a marker of both numbers, though he considers the opposition of *cat–cats* to be privative.[4] Hence J. Cantineau, in this case, does not differentiate the opposition of the plane of expression and the plane of content. However, an opposition is an event of the plane of content, with the meaning serving as the basis for comparison of its members, although any marked *DF* of the opposition correlates with a definite formal exponent.

Still, we should stress the prevailing desire of linguists to accept the privative nature of any opposition of the type *A*–non-*A*. Sometimes it creates artificiality in the interpretation of the relations in the system, as, for example, the definition by M. Ruipérez of the Greek presence as a marked member and the aorist as an unmarked one,[5] while it is known that both aspects possess formal signs and independent meaning. Here equipollency of the relations of Greek aspects is evident.

To admit the exclusively privative nature of any opposition is tantamount to proclaiming the uniformity of all possible relations in the system. However, it would be logical to expect that the relations in a paradigmatic system are not absolutely homogeneous, but in a known measure complex, though not too varicd (as in syntax or syntagmatics).

In detecting oppositions in a system, the criterion for joining any two grammemes into an opposition is the link between the contrasted units. Oppositions reveal both paradigmatic and syntagmatic relations between the grammemes. Elements with no basis for comparison cannot enter the same binary opposition and cannot be connected by certain uniform grammatical relations.

[3] Trnka (1961) 134.
[4] Cantineau (1952) 28-31.
[5] Ruipérez (1953) 241-252.

Thus, in the English verb system the past indefinite *I wrote* is not contrasted to the present perfect *I have written*, since they have no basis for comparison; they belong to different temporal planes and to different stages of temporal relation — the indefinite entering the sphere of simultaneity, and the perfect the sphere of anteriority. The members of different oppositions, in similar cases, may be contiguous to each other in use; this is due not to their belonging to one opposition, but to the intersecting of the spheres of the tense proper (or, in other words, time) which marks the axes of orientation, and the temporal relativity (or perfectness) which marks the relations of the verb forms WITHIN the bounds of one axis of orientation — the present, past etc., expressing either simultaneity with the axis or priority to it. That members of an opposition should necessarily belong to one common sphere is stressed by L. R. Zinder in his article "On contrasts in a language system". His ideas on the subject may be summarized in the following: we may contrast events and concepts, relating to the same sphere and linked between themselves. The contrasts form a system. Thus, the notion of a system came to be used together with the notion of contrast. L. R. Zinder speaks of the importance of contrasts in a grammatical system: no grammatical categories are possible without contrasts, which in grammar can bring about more unexpected and interesting results than in phonology.[6] One cannot but agree with such considerations. Another example: the grammemes *am writing* and *have written* have as a basis for comparison the *DF* of the axis of orientation of the action, i.e. the present. However, they find themselves neither in a privative nor an equipollent opposition, since each of the grammemes separately possesses one different strong (marked) *DF* — duration and perfectness, having only one temporal *DF* serving as a common ground for comparison. Such a correlation for the members of a contrasted pair within the verb system is unusual, which is confirmed on the syntagmatic axis: in no contextual conditions can one form substitute another and preserve the identical communicative trend of the sentence.

[6] Zinder (1962) 121-122.

An important criterion for detecting oppositions expressing a grammatical category is the behavior of pairs of grammemes in relation to neutralization. The process of neutralization will be considered in detail in the next chapter. Now we shall but point out that the content of the strong member of a privative INTRAPARADIGMATIC opposition may be neutralized; in it both members are linked by an essential (for the given opposition) grammatical feature (or features) which serves as ground for comparison, i.e. where the members are isomorphous, as in the case of the opposition of the English indefinite *(I write)* and continuous *(I am writing)*. In such a case one counter-member of the opposition marks a strong *DF* (duration, here) and the other leaves this *DF* unexpressed. Such a kind of contrast corresponds, with great precision, to the logical definition of a privative opposition, the more so that it is subject to neutralization, a fact which is also a strong confirmation of the correctness of the definition suggested. The meanings of the forms which are not in an opposition are not neutralized, but can only adjoin in certain similar contexts, which is not always the case.

The members of an equipollent opposition are neutralized on a different basis, and the limits of the opposition may be defined again by its behaviour in contextual conditions of neutralization. Of the temporal forms, for example, past and future cannot be neutralized and, hence, cannot be contiguous to each other or be used instead of each other in speech, because they are not in opposition; the present may be in the position of neutralization, both of the past in the cases of the so called 'historic present' and of the future in contexts where adverbial modifiers of future constitute the environment of the form, since it is the initial member of the two oppositions, both in pair with the past and in pair with the future (see chapter 11).

Thus, the boundaries and composition of oppositions are verified not only by the logical relation between members when there is a basis for comparison, but also in their relation to neutralization, i.e. on the syntagmatic axis. This is natural since the relevance of an opposition is displayed both on a paradigmatic and syntagmatic axis. An opposition functions in speech and is realized in a certain distribution. The contrast of uses in speech is the function of an opposition, and the meaning of the members of an opposition is precisely the function of their use.[7] All these criteria can be conveniently illustrated by the manifold relations among members of paradigms of the English verb.

Besides the above two types of oppositions, privative and equipollent, the English verb system has another type, which in the relation between the members holds a kind of intermediate position between privative and equipollent oppositions. Here we mean the contrast of grammemes denoting simultaneity with the axis of orientation and grammemes denoting anteriority to it, i.e. the opposition of the non-perfect and perfect forms of all temporal planes. This type of opposition, as is known, is characteristic of many European languages. Simultaneity is not marked formally, though in regard to anteriority it holds a different position than does an indefinite form in a privative opposition to a continuous one. An indefinite form simply leaves unexpressed the *DF* affirmed in the continuous set, whereas the forms denoting simultaneity with the given temporal plane do not simply leave anteriority unexpressed, but themselves carry information about the *DF* of simultaneity with the temporal plane, in contrast to the *DF*, produced by the perfect set. Hence, both members of the opposition express features relevant for the given opposition, because the opposition itself is based on comparing the event taking place simultaneously with the axis of orientation or preceding it. From the point of view of the plane of expression, simultaneity is not marked at all and is a privative member, while from the point of view of the opposition itself, i.e. of the plane of content, simultaneity is one of the contrasted features and is logically equal to anteriority. All this confirms our assertion that this opposition holds an intermediate place between a privative and equipollent one; it can be conditionally called privative-equipollent. In any case, no matter how expedient may be the introduction of such a term for this

[7] A similar view is expressed in: Tellier (1962) 16.

kind of opposition, its specific property confirms the postulate that logical relations between the members of morphological oppositions cannot be reduced to privative ones only, and that relations between grammatical units may be varied, the latter also affecting the behaviour of the members of an opposition in the process of neutralization. Compared with privative, this opposition behaves differently when neutralized, since its members are linked in their content by equipollent relations. The neutralization of the meaning *(DF)* of anteriority and its expression in the forms of the sphere of simultaneity is a rare phenomenon and depends on rather limited conditions of distribution. The basis for comparing the members of this opposition is the general temporal plane of the given axis of orientation. Through the nature of meaningful relations between its members this opposition may be compared to the opposition of singular–plural in the noun. Even if we accept the definition of the essence of grammatical number, given by A. V. Isačenko, viz., that number denotes the *DF* of expressed (plural) and unexpressed (singular) articulateness[8] (though A. V. Isačenko proceeds exclusively from the privative nature of opposition), we may state that in their relation to each other, both the features of simultaneity and priority (anteriority) are equally relevant in the plane of content and hence, the members of the opposition are equivalent. Such an opposition does not expose an asymmetry of the plane of content, as in the case of privative oppositions where the extensive member is but a formal, initial point of the opposition, containing the basis for comparison and carrying no information about the contrasted *DF* marked in the strong countermember. In the singular — plural opposition, the singular contains information precisely about the number even though, formally, this *DF* may have zero exponent in some languages.

In any case, the peculiarity of the correlation of members of the above oppositions on the paradigmatic axis seems evident. It is also reflected in syntagmatics, first of all in neutralization.

The morphological system, despite its general order and relative symmetry and proportion, is a complex mechanism, and it is

[8] Isačenko (1961) 36 ff.

difficult to suppose that all relations in it may be reduced to some uniform nature of ties. Between the members of an opposition there exist hierarchical relations, defined according to the nature of the grammatical category expressed in its units and joined into oppositions. We may expect, therefore, that the nature of meaningful ties in different oppositions may be varied, depending on the difference in grammatical categories.

Since oppositions are based upon both the difference and the community of the grammatical meaning of their members, the features of similarity and difference should form the basis of determining the nature of oppositions. But they are not always easy to define. That is why N. S. Trubetzkoy says that the determination of oppositions depends upon the point of view.[9]

Since, by the nature of logical relation between their members, morphological oppositions are not too heterogeneous, as there exists few morphological types generally, it stands to reason to suppose that oppositions should be proportional, i.e. they should recur in similar sets and microsystems.[10]

If to illustrate oppositional analysis we take the paradigmatic system of the English verb, we shall see that it is linked by intra- and interparadigmatic relations, realized through oppositions, which represent intersections of binary ties, regulating and penetrating the whole verb system. Any verb microsystem (of the passive voice, verbals) has the number of members divided by two and falls into binary clusters which intersect and mix among themselves. In studying oppositions as symbols of relations, created by the generalized content of the grammatical category expressed in the given opposition, one should keep in mind only the invariant meaning of every member of the opposition as it exists in the language system. The actualization of these relations in speech, on a syntagmatic axis, is a question of special consideration. The first link in a series of rules on the actualization of oppositions on the

[9] Trubetzkoy (1936b). A detailed description of manifold relations and correlations in the structure of language is given in the book Brondal (1943).
[10] Cf. Cantineau (1952). Kurylowicz (1956) 21 speaks of proportional relations and polarity as inherent laws of the system of language.

syntagmatic axis, in the sentence, is the process of neutralization, which is a link between the paradigmatics and syntagmatics of the members of the opposition.

7.

NEUTRALIZATION OF OPPOSITIONS

One of the most interesting processes, discovered during the last two decades due to the functional structural analysis of language events, is the neutralization of oppositions. The idea of neutralization in regard to phonology was first advanced by N. S. Trubetzkoy and A. Martinet at the fourth Congress of Linguists in 1936,[1] and it found response among scientists of many countries. But it was much later that this process was discovered in units other than the phoneme. Suffice it to say, that it was neutralization, mainly in morphology, that was the subject of the whole volume of proceedings of the Paris Institute of Linguistics, which contains forty-five often contradictory points of view.[2] In the volume, the most varied examples of neutralization are cited, such as: suspending distinctions of voicelessness — sonority in Russian *kod–kot*, suspension of contrasts of number (person) in French *il mange–ils mangent*, of aspect in the Russian future *ja budu pisat'*, of nominative and accusative cases in neuter gender in languages with a four-case system, of singular and plural in a number of nouns of the type '*pluralia tantum*', etc.; hence, according to the above examples, the idea of neutralization is used by different authors to explain different phenomena.

The process of neutralization is elaborated in greater detail by

1 Trubetzkoy (1936a) 29-45; Martinet (1936) 46-57.
2 *Travaux* (1957); see also Hjelmslev (1939) 51-57; Bazell (1949) 77-86; Hjelmslev (1968) 121 ff.

N. S. Trubetzkoy in regard to phonemes.[3] The problem of neutralization proves to be more complicated in morphology, which deals with bilateral signs having both expression and content. In phonology we have the neutralization of a kind of 'meaning' of a phoneme, i.e. of one of its *DF*, which make up its sense-distinguishing essence; thus, at the end of some words, like Russian *rod–rot*, etc., the distinction in sonority is suspended.

If there are oppositions in a morphological system, there is a possibility of neutralizing them in certain particular conditions, which introduce a correction into the systemic content of the members of oppositions. The isomorphism of the levels of the language structure makes it possible to use events of phonology as a standard for finding similar relations at another level, considering, of course, the difference between the unilateral nature of the phonological sign and the bilateral nature of signs which belong to a higher level of language structure. It remains to elucidate what we are to understand by neutralization — a paradigmatic or syntagmatic event, permanently existing in the system of signs or occasional, changeable, caused by definite conditions of the functioning of the sign.

One should be rather careful in transferring the notion of archiphoneme to the higher levels of the language, since the bilateral nature of any sign, except the phoneme, hampers the selection of a unit combining the features of various members of an opposition. But in principle, the search for an archi-unit reflecting features common to both members of an opposition may be justified, though such a unit cannot combine in itself features of both the plane of content and of expression, due to the bilateral nature of the sign.

To solve the problem of neutralization we should, first of all, distinguish neutralization in general, as a process, from neutralization of sense-distinguishing contrasts (oppositions). Neutralization may occur, of the notion of singular in abstract nouns, or in distinguishing the aspect in future tense in a number of languages;

3 Trubetzkoy (1939) 206-217.

also, a certain seme of a word in a given context may be neutralized (as in the example, cited by V. V. Ivanov in regard to neutralization of the seme of 'animal' in the specific use of the word 'an ass').[4] In such cases there is no opposition. A different situation is the neutralization of oppositions.

Among the cases of neutralization of oppositions, events, different in their nature, should be distinguished. First of all there exists PARADIGMATIC NEUTRALIZATION, an example of which is neutralization of the opposition of accusative and nominative in nouns of neuter gender; it is a permanent neutralization, inherent for the given type of paradigm, in which the contrast realized in parallel paradigms of the declension of masculine and feminine nouns is not actualized. In SYNTAGMATIC NEUTRALIZATION, the contrast is suspended in speech under certain distributional conditions, as, for example, the suspension of the distinction between voicelessness and sonority in the ending of some words (type *kod–kot*), with the members of the phonemic opposition keeping their sense-distinguishing features in the paradigm. So, if we take neutralization for a syntagmatic, temporal, variable phenomenon (and not constant, or peculiar for the language and the given opposition), if we consider it as a sign of the interaction of the paradigmatic and syntagmatic axes of the language, then this event will never cover such cases as *pluralia tantum* and the differentiation, permanently fixed in the language, of the use of cases dependent on the semantics of the governing verb.

Synonymy and homonymy are examples of constant neutralization; they are dealt with in the article by O. S. Akhmanova, published in the cited volume of *TIL*.[5] O. S. Akhmanova thinks that we can speak of neutralization in bilateral units only in the plane of homonymy (neutralization of expression) and synonymy (neutralization of content). These events are also fixed in the language structure at a certain historical period of its functioning; they are not variable categories in the synchronic plane and do

[4] Ivanov, the answer to the questionnaire offered by A. Martinet, *Travaux* (1957) 45-48.

[5] O. Akhmanova, the answer to the same questionnaire, *Travaux* (1957) 15-17.

not imply the presence of an opposition. Neutralization is often confused with various other events, e.g., syncretism and free variation (facultative change). As is known, L. Hjelmslev in his *Prolegomena* did not distinguish neutralization and syncretism,[6] but in an article, especially devoted to neutralization,[7] these notions are distinguished. A clear example of syncretism is the historical merging of several cases into one (e.g., instrumental and local into dative); other instances of the merging of potentially distinguishable grammatical meanings in one form are second person singular and plural in the English *you are*, absence of any distinction of gender in the plural of German declensions, of aspect in the Russian periphrastic future, of the praesens and aorist in the Greek future, etc. Such examples are often cited as cases of neutralization. However, this is a qualitatively different event, since in the given class of units in the given language such oppositions do not exist, while only those oppositions are neutralized which actually exist in a given system and are not only theoretically possible in other languages. Neutralization implies the presence of a distinction, while syncretism testifies to an identity, to the existence of one unit only and not two, since in syncretism potentially different meanings are merged into one and the same form. Syncretism is related to polysemy; it seems to be an historically conditioned polysemy or a polysemy to be discovered by means of comparison with other languages.

As free variation may be considered instances of the use of various prepositions and conjunctions to express identical relations of different cases in some combinations of the type of the Russian *čaja*, *čaju*, since here we have no oppositions.

Thus, we should distinguish neutralization from the partially similar, but qualitatively different events of homonymy, synonymy, syncretism and free variation.[8]

Controversial cases may be found in the usually cited examples of paradigmatic neutralization; it may be noted only when the

[6] Hjelmslev (1968) 121 ff.
[7] Hjelmslev (1939) 55.
[8] On the differentiation of all these notions see Khlebnikova (1970).

opposition being neutralized is supported in an analogous class of units in an adjacent paradigmatic scheme.

Such a phenomenon as absence of gender in the plural of German declensions does not come under the notion of neutralization, since that system of declension contains no such differentiation at all and, so, there is nothing to be neutralized; neither is there any opposition of praesens — aorist in the Greek future. A doubtful case of paradigmatic neutralization is the absence of any distinction of nominative and accusative cases in nouns of neuter gender, since these cases are not distinguished under any conditions.

A case of permanent (and not syntagmatically conditioned) paradigmatic neutralization of the *signifiant* (plane of expression) may be the non-distinguishing of number in some denominations of animals in English like *sheep*. Here a member marked by the sign of plurality is neutralized, and there remains in the paradigm a member unmarked by this sign, and hence capable of including in itself the notion of plurality. Neutralization can hardly be fruitfully considered as a diachronic process leading to different results. Neutralization is a synchronic phenomenon concerning oppositions, and an opposition is a synchronic event as well, since the nature and composition of oppositions can radically change during the history of the language development. The notion of neutralization does not include cases when some subclasses of adjectives have no opposition in degrees of comparison, for it is not possible to speak of neutralizing oppositions when there is no opposition at all. Thus, for example, a change in the degrees of comparison may be possible mainly in adjectives which form contrasting pairs like *big–small*, *narrow–wide*, *bad–good*, etc. To the neutralization of number may be referred cases cited by B Trnka, like *I have no son–I have no sons* and also cases when the purpose of the number is use for the purpose of concord, as *my brothers are merchants.*[9] If we admit that neutralization is a phenomenon of syntagmatic order, then we cannot give as examples of it forms and their usage, fixed in the language.

[9] Trnka (1961) 137.

We can speak of neutralization, if in the system of a class of signs there really exists a given opposition, joining two members of the system by certain *DF* and possessing a sufficient degree of reproductivity and abstraction, and also when this opposition is realized, at least in one context. Oppositions, and hence, their neutralization, are possible only in regard to phonology and morphology, i.e. where there exists a paradigmatic of *DF* and categories. As far as syntax is concerned, it contains paradigmatics of models and classes of combinations, but not polar correlations, all the more so that the main category of syntax, the nature of syntactic relations, is not covered by any paradigmatics. But this is a special, complex problem, and it is very likely that close investigation of syntactic phenomena on the basis of invariant meaning and *DF* may lead to the discovery, in this sphere as well, of some relations which can be explained by means of oppositions.

Hence, we can consider oppositions, as well as neutralization, to be a phenomenon, peculiar to paradigmatic sets, the neutralization reflecting the interconnection between the paradigmatic and syntagmatic axes of the language. Regular oppositions of definite grammatical *DF*, existing in the language as a system, as an inventory of units and rules of their organization, are realized on the syntagmatic axis in speech situations, in context, which can create conditions where one of the *DF* of the opposition may prove to be suspended, irrelevant in the given context. Thus, in these sentences — *While she watched TV, I slept*; *while she was watching TV, I was sleeping*; *while she was watching TV, I slept* — the *DF* of the duration of the action which can be actualized in such a context may become irrelevant in a position where there are two simultaneous past actions. Then the paradigmatic contrast of non-duration–duration is suspended, and the unmarked member appears in the position of neutralization, the one which possesses a more general meaning and itself expresses a *DF* common for both members of the opposition. Common for the indefinite and continuous sets are two *DF* indicating the simultaneity of the action with the present moment. These two *DF* also enter the grammatical (categorial) content of continuous forms marked, in addition, in the *DF* of duration. Such contextual situations seem to shorten the system of categorial oppositions forming 'an individual system', valid for certain contexts, which is mentioned by N. S. Trubetzkoy in regard to phonemic neutralization.[10]

Thus, not in all cases, where a durative action is to be expressed in English, should a continuous set be used, since, as a marked member of a neutralizable opposition, duration can be neutralized, and then, according to the rules of neutralization, an unmarked member necessarily finds itself in the position of neutralization, having a more general meaning and possessing a neutralized *DF* (duration in this case), which is not marked.

We may consider as neutralization of duration those cases when the present indefinite is used instead of the present continuous in describing a sequence of actions, as *he goes to the blackboard, takes the chalk, now he writes.* A similar neutralization, but combined with an adverbial modifier, is in the example: *I have worked here many years.* In the position of neutralization here again is the unmarked member, which expresses a common *DF* for both members of the opposition — anteriority to the present. The same content may, of course, be expressed by the perfect continuous form *I have been working here many years,* where we have a '*Relevanstellung*',[11] i.e. a situation, in which the sign of duration is realized.

So, the neutralization of oppositions is a phenomenon of a syntagmatic order; it is a breach of the symmetry of systemic contrasts in certain conditions of the context, a kind of natural exception from the rules of oppositions, a contracting of a categorial opposition consisting of two members, into one member, one grammeme, possessing semes which are active for both members of the oppositions.

The very existence of the process of neutralization of one member of an opposition, formed by grammemes of the given language is, perhaps, due to the general trend of the language to do away

10 Trubetzkoy (1939) 206-17.
11 Trubetzkoy (1936a) 30.

with redundancy of grammatical units, a tendency which has appeared in the process of the historical development of paradigmatic sets. All this is also connected with and conditioned by the principle of least effort, which to a great extent regulates language processes.

The specific nature of morphological neutralization, syntagmatic in its basis, consists mainly in that not the form is neutralized, but the meaning or, rather, one component of the meaning, relevant for the given opposition, and on which the opposition is built. It cannot be otherwise, if we speak of neutralization not generally, but of neutralization of oppositions, since an opposition is an event of the plane of content of a grammatical category, which is supported but not conditioned by the form. That is why it is hardly expedient to introduce the notion of an archimorpheme. i.e. some form under an asterisk capable of combining the relevant features of the opposition and reflecting the community of the formal features of the members of the opposition subject to neutralization. If we are to select a certain unit, reflecting the community of the features of opposed members, i.e. their basis for comparison, such a unit may be an archisemanteme (or archisememe),[12] and not an archimorpheme, since it may be a question of the general features of the CONTENT and not the form.

Neutralization is a temporal phenomenon, and it is always partial. There can be no complete neutralization, since that would mean complete absence of an opposition and is equal to synonymy or syncretism.

Thus, neutralization is impossible when there is no given opposition in the given language. If in English there is no opposition of second person singular and plural, then there can be no neutralization of number. (Examples of neutralization, however, are often cited precisely where opposition does not exist). This is an essential condition of neutralization. Neglecting this condition obliterates the boundary between neutralization and qualitatively different events and abolishes this specific process. The opposition itself is

[12] Cf. the use of this term by Ruipérez in *Travaux* (1957) 118; Khlebnikova (1964b) 63.

recognized through neutralization. Thus, the possibility of suspending the relevant feature of duration and substituting it with a negative (privative) member, testifies to the fact that the grammeme, expressing duration on the basis of simultaneity with the moment of speech *(I am writing)*, and the grammeme expressing only simultaneity with the moment of speech *(I write)* are in a categorial opposition by aspect, in which one positive member marks this *DF* and the other leaves it unmarked, both members keeping those components of the meaning, which serve as a basis for comparison and as the foundation of the opposition, i.e., in this case, the semes of the temporal axis and simultaneity. The feature or features common for the members of the opposition, and not distinguishing them, remain relevant in neutralization.

Neutralization also stresses the importance of distinguishing the invariant and variant meaning of categorial forms, since the invariant meaning present in all the uses of the form is neutralized, and the contextually or lexically stipulated variants are not.

It follows from the above that first those oppositions may be neutralized, among whose members the relations are isomorphous, i.e. having points of contiguity, and not polar. If we understand polarity in its broad sense as any contrast, then any opposition will be polar, and any relations in the language are also polar. But in some oppositions (in equipollent, in particular), the content of the members is actually contrasted, both members asserting *DF* incompatible in usual context situations. For example, the singular is incompatible with the plural, or the present with the future in a minimal neutral context, such as **The boys is*; in a situation where a present action is meant — *I know English* — one would not use the future. In privative oppositions, since one member affirms the *DF* lacking in the other, the basis for joining the two units into one opposition is some other essential *DF*, common to both members and remaining in the position of neutralization. It is exactly such a substratum of the opposition in the form of a general basis for comparison, affirming the relevant *DF* for both members of a privative opposition, that provides the neutralization of the opposition, because this common *DF* turns the members of the opposition into isomorphous ones, as in the opposition *I have written–I have been writing*, where in both forms two *DF* remain common — those of temporal axis and of perfectness.

Neutralization should not be identified with synonymy, mainly because a synonym is a constant event fixing some sign in the synonymic set, while neutralization is a syntagmatic event, creating no synonymy but temporarily (contextually) bringing two members of the paradigmatic set closer together, on the basis of a categorial *DF* common to them.

Hence, it is possible to assert that privative oppositions are neutralizable. But we consider as privative interparadigmatic oppositions as well. Can interparadigmatic oppositions also be neutralized? The case is complicated by the fact that every member of such an opposition is multinuclear, so that neutralization is possible in every separate case, in regard to a separate paradigmatic set or type.

As to oppositions built on the contrast of incompatible *DF*, their neutralization has its peculiarities. Equipollent oppositions are built on contrary conceptions (and not contradictory, as in the case with privative oppositions). The difference between contrary and contradictory conceptions in logic is dealt with by A. V. Isačenko in the above article on binarism.[13] But he thinks that all oppositions, built on contrary conceptions, may be reduced to privative relations based on the presence or absence of a definite *DF*. That this is not so is proved by the relation to neutralization of those oppositions which we consider to be equivalent in the semantic content of their members.

N. S. Trubetzkoy pointed out that neutralization of equipollent oppositions is a rare case.[14] At the morphological level the same may be observed. Equipollent oppositions refer to permanent rather than to neutralized oppositions, since both members possess relevant signs, and the basis for comparison is a more general categorial meaning, lying at the base of a number of other oppo-

13 Isačenko (1963) 41.
14 Trubetzkoy (1939) 74.

sitions, and not the meaning specifically inherent for the given opposition. Thus, the present and past *I see a car* — *I saw a car* or the indefinite and perfect forms *I see* — *I have seen* are not interchangeable in a usual, stylistically neutral context, since for the first pair an abstract distinction of action in regard to the moment of speech will be common, and for the second pair, the plane of the present will be common, common also for the oppositions in aspect within the same temporal plane. If in the above privative oppositions both members are joined on the basis of TWO fundamental sense components of contrasted members — simultaneity plus temporal plane, or anteriority plus temporal plane, i.e. common *DF* are expressed both by a more general and more individual feature — then in equipollent oppositions only one more general *DF* remains as the basis for comparison. From this stems the limitation in the process of neutralization, since neutralization is a situational, partial contraction of the paradigmatic set, and it is evident that the contraction of members possessing equally relevant individual *DF* would be unnatural.

In the position of neutralization of privative oppositions, there always appears the unmarked member, both in form and content, and the reverse is impossible, i.e. a marked member cannot be used instead of an unmarked one, since the process of neutralization is irreversible (e.g. the past indefinite may be used in the sense of the past continuous, while the reverse is impossible; one cannot say *I was coming yesterday* meaning *I came*). The question arises, then, of which member may appear in the position of neutralization of equipollent oppositions?

Neutralization of members of an equipollent opposition in time takes place through TRANSPOSITION, i.e. provided the temporal form is displaced into the temporal plane unusual for it; the direction of transposition and, hence, neutralization shows the composition of the oppositions in time (tense). To illustrate: a typical case of such a transposition is the use of the 'historic present'. The past action is displaced into the plane of the present, and so the present expresses a temporal plane unusual for it. The member which is the initial point of the opposition finds itself in the position of neutralization. The present may also be displaced into the plane of the future, provided a corresponding adverbial modifier is present in the immediate environment of the form, as in *I leave for Moscow tomorrow*; here again the present is possible in the position of neutralization of the opposition 'present–future'. The reverse is not typical, i.e. the future or the past are not used instead of the present. The past reflects a retrospective direction from the present, and the future a prospective one. Both tenses are linked with the present, which is confirmed by the use of the present in the position of neutralization. The past is not contrasted to the future, so that the one does not substitute the other in the position of neutralization. However, we deal here with neutralization of equipollent oppositions, in which both members possess relevant *DF*; therefore, it may be expected that in some rare cases of neutralizing their members, the appearance of a stronger member marking the grammatical *DF* in the plane of expression is not excluded. Thus, such sentences as you *will be Mr. X.*? meaning *you are* are possible, though rare, where we have the neutralization of the opposition 'present–future', and in the position of neutralization remains the stronger, marked member. The above confirms the different nature of such oppositions compared with neutralizable privative ones, like *I have written–I have been writing*, where in the meaning of the perfect, the marked, i.e. the perfect continuous member, is impossible (one cannot expect a sentence like **I have been finding a purse* instead of *I have found a purse*).

In privative-equipollent oppositions neutralization is also a rare event. But here as well, the correlation of members of the opposition in speech directly depends on the peculiarities of the temporal plane, on the axis of which the opposition is based. For example, on the axis of orientation of the past, due to close syntagmatic connection of all forms of the sphere of the past (the rule of sequence of tenses), the members of this opposition may draw closer, and the past indefinite may sometimes be used instead of the past perfect; in the sentence: *he knew I wrote that book two years ago* anteriority is neutralized. In such colloquial expressions as *I forget your name*, *I hear you are a painter*, the *DF* of anteriority is also neutralized, and the member unmarked by this *DF* is used in the position of neutralization.

Still it is noteworthy that for the latter two types of oppositions neutralization is not natural, precisely because they are not privative oppositions, and both members affirm, though in a different degree, a *DF* relevant for both members of the given opposition. Hence, both members are essential for the system of oppositions and are always present in it. This principle will be confirmed by some concrete examples in the next chapters.

We should keep in mind, however, that members of non-privative oppositions cannot be absolutely equivalent in value, since they would then not be in opposition, which always implies a definite hierarchy of members, an asymmetry of interdependent grammatical forms.[15] Thus, if in privative oppositions we can speak of a weak, unmarked member and of a strong, marked one, then in equipollent ones there is a weaker and a stronger (more specified) member of the opposition.

A. V. Isačenko calls the perfective aspect a more specified member of the opposition in aspect.[16] R. Jakobson introduces the terms 'less specified unmarked member' and, vice versa, 'more specified marked'.[17]

Though both authors are referring to members of privative oppositions, since they reduce all oppositions to privative, actually these denominations will suit oppositions with equivalent relevant members. Such a weaker member, with a more general meaning will be, e.g., the present in temporal contrasts, or the singular in opposition by grammatical number, from which it follows that in the position of neutralization precisely this weaker, less specified member will appear. Such members are initial in oppositions and themselves serve as a kind of basis for comparison with other contrasted units. Thus, temporal oppositions are considered in regard to the moment of speech, proceeding from it, and anteriority is relevant in regard to simultaneity, only when compared with it.

15 Cf. Jakobson (1932) 79 ff.
16 Isačenko (1960) 133.
17 Jakobson (1959) 140.

Everything said above confirms the idea that the relation to neutralization helps to detect the oppositions themselves and define their nature.

The privative nature of oppositions in aspect and their ability to be neutralized are confirmed by another factor: the marked strong member of such oppositions may be easily omitted from the verb system without lessening its ability to express durative actions through weak members. Thus, in the microsystem of the passive voice some durative forms are completely absent, though it does not create difficulties in expressing corresponding actions. In the sentence *I have been engaged in this conversation for quite a time already*, the meaning of the durative action, approaching closely the present moment or continuing at present, is clearly enough expressed in the perfect form, unmarked in the *DF* of duration. The absence of the future continuous creates no obstacles in expressing durative actions in the passive as in: *the lecture will be delivered soon.* In the active, a corresponding sentence may be expressed in the form of the future continuous: *he will be delivering his lecture soon.*

Neutralized members of privative oppositions create redundant forms in the verb system, since it is such members that can be dropped due to the ability of the unmarked member to express a common meaning, inherent for the neutralized member as well. It is known that the lowest statistical potential is characteristic of such grammemes as the future continuous and the perfect continuous, as well as of a number of other forms.

B. Trnka draws attention to the fact that not every verb enters the opposition of indefinite-continuous forms, but every verb enters the opposition 'present–past'.[18]

It follows from the present discussion that the nature of oppositions, their inner qualities, and correlation of their members are specified on the syntagmatic axis and first of all in the process of their neutralization, because neutralization is a syntagmatic event. The concept accepted in linguistics, that an UNMARKED member

[18] Trnka (1961) 133.

appears in the position of neutralization, has a deep meaning and confirms the irreversibility of the process of neutralization, though the degree of unmarkedness of the member of an opposition characterized by a more general content is different depending on the nature of logical relations between members of the opposition, on whether it is privative or otherwise.

In the process of neutralization, one *DF* of the strong member of the privative opposition is lost during the preservation of other *DF* of the given opposition remaining as grounds for comparison. Thus, in neutralizing sonority in the opposition *d–t*, sonority is lost but other features are kept, such as occlusion, dentality etc. In privative morphological oppositions the seme, which marks the strong member, is lost in neutralization, e.g., the seme of duration in the indefinite-continuous opposition, while other *DF* of the opposition — temporal relativity and temporal plane are preserved. In equipollent oppositions, the categorial plane of a stronger member is also suspended; however, since the neutral member does not retain any component of meaning, which would be characteristic only of the members of the given opposition (and comparison of members may take place only on a basis common for other oppositions as well), neutralization of such a member takes place through transposition, i.e. by transferring the weak member, in the position of neutralization, completely into the sphere of the strong member under neutralization. Thus, in the sentence expressing narration in the past, the present, displaced into the sphere of the speech event *(suddenly he sees a car, rushes to it)*, is completely transposed into the sphere of the past and vice versa, the past into the sphere of the present, since, due to the equipollency of the opposition, one member cannot express even a part of the *DF* inherent for the other member. That is why we say that non-privative oppositions are neutralized through transposition, and transposition more often than not implies a shift in the communicative purpose of the utterance, a change in the stylistical colouring of the sentence.

Hence, neutralization of equipollent oppositions is a rare phenomenon. Investigating the action of the law of neutralization in diverse situations found in the text and individual speech is an essential task of syntagmatic analysis, i.e. the analysis of the use in the sentence (speech) of members of morphological paradigmatic sets.

8.

INTERACTION OF FORM (PLANE OF EXPRESSION) AND MEANING (PLANE OF CONTENT)

Though the plane of expression and the plane of content are each a special subject of study by different methods, a permanent link and interaction are established between them — expression is reflected in content, and the content in the expression. The morphological structure is defined, proceeding from the purpose of the morphemes, and from their categorial content. The question of the analytical form is also solved, proceeding from the content of its components.

Homonymy is an event of the plane of expression, fully defined through the plane of content, through distributive properties of forms, caused, in the final count, by their content. Homonymy is closely linked with polysemy, which is characteristic of the content of some morphological units.

Interesting examples of the interaction of form and meaning are oppositions in verb microsystems. Though oppositions as a means of realizing a grammatical category are an event of the plane of content, still, all marked strong *DF* of oppositions are directly reflected in forms — carriers of these *DF*. This reflection stems from the inventory of formal elements existing in the given microsystem of the given language. Different *DF* of the meaning of forms find reflection in only an insignificant number of morphemes capable of entering varied combinations among themselves. In any paradigmatic system there is some initial form serving as a standard for comparison and for further formbuilding, which is similar to the situation existing within oppositions based on content, where the basis for comparison of grammemes is contained in the initial counter-member. It is interesting to note that the basic form (such as, for instance, the base of the verb, as in *love-ed, -ing, will love*) often becomes the starting point of meaningful oppositions.

Such an initial form among inflexional formants of the verb serves as its base, being a component of the present, which serves as an initial point of comparison both in the plane of expression and in the plane of content.

Any complication of this initial form by inflexional formants leads correspondingly to a change in the categorial meaning of the grammeme. The change in the categorial content of the base may be due to homonymy as well as, e.g., in the case of the imperative or the remnant of the Old English conjunctive *(I go, go!, if he go)*.

Thus, the past indefinite is created by adding a corresponding suffix (in the regular type *lov-ed*), and is simultaneously included into the opposition of time in the plane of retrospective relation to the moment of speech.

The modal nature of the future tense calls for the use of verbs, modal in origin, as formants of a prospective opposition in a language where there is no special suffix formant for the future tense. The sense (lexical) part of the analytical form (as distinct from its grammatical part) is reflected in the stem form of the infinitive *(will go)*.

Duration is conditionally reflected in another combination, most closely showing a durative process — the auxiliary inflexional formant *be* with participle I, which in its turn is also built with the help of a suffix added to the initial non-finite form of the infinitive, *was writ-ing*. Different temporal variants of the continuous form are also reflected in the grammeme of the durative aspect by changing the tense of the auxiliary verb.

Perfectness is reflected in elements most suitable for the purpose — the auxiliary *have* and the notional verb, taking the form of participle II. The combination of perfectness and duration also finds reflection in the corresponding combination of formants of both perfectness and duration, while the passive uses formants already functioning in the scheme of tense-aspect, but in a different combination, which again reflects the content of every given form-type *(has been developed)*. The specific nature of the modal content of the subjunctive mood and the limitation in the inflections of the English verb gives rise to homonymy within the microsystem of the subjunctive[1] *(would have spoken, if he had spoken)*. The development of homonymy in grammemes expressing an unreal action was a regular outcome of the loss of corresponding inflections, because if the grammatical category of the subjunctive and/or conditional (whichever term may be used to denote it) were expressed by means of any formative not present in the basic microsystem of tense-aspect, it would contradict one of the structural principles of the organization of paradigmatic schemes of the verb, viz., THE STRUCTURAL UNIFORMITY of inflexional morphemes, the standard for which is the initial microsystem of tense-aspect.

Although in all microsystems of the verb there is the same narrow range of inflexional formants (in various combinations), which are to be found in the basic microsystem of tense-aspect, the nature of grammatical categories, reflected by these morphemes, is different in various microsystems. The diversity of the formal side of an opposition is ensured either due to the varying of the combinations of formants or by using the forms already present in the basic microsystem as homonyms to express another grammatical category. Then the differentiation of homonymous forms is completely displaced into the field of the context, and their categorial content is learned from their distribution. Thus, the plane of content of a formal unit is revealed in two aspects: through its paradigmatic relations as a member of the opposition by the grammatical *DF*, and through the whole sum of environments natural for the given form as a unit of a definite grammatical meaning. Both these sides of the content make up what may be called the structural meaning of the form. The fact that the para-

[1] On the problem of mood see Khlebnikova (1965b).

digmatic content of the grammeme interacts with its distribution manifests itself, among other factors, in the process of the neutralization of oppositions.

It follows from the above remarks, that the analysis of a morphological system and, in particular, the verb system, implies taking into account the permanent interaction of the form, content, and distribution in speech.

But it does not mean that all these three factors are in a symmetrical connection and interdependence. The maximum asymmetry is found in the correlation between expression and content, this being reflected first of all in the presence of homonymous and synonymous sets, when one expression may have several contents, and vice versa. This linguistic fact should be taken into account in analysing language events. Neutralization (syncretism) also testifies to asymmetry in the plane of expression and the plane of content, the lack of their full equivalence.[2]

[2] The asymmetry of linguistic units of different levels is discussed in: Karcevskij (1964) 81-87; Bazell (1945-1949) 139-145; Jakobson (1932) 79.

9.

THE NOTIONS OF MICROSYSTEM, MACROSYSTEM, AND SYSTEM

It follows from the above that morphological (grammatical) categories should be studied and described on the basis of oppositions, taking into account all that is connected with oppositions, in other words, such events and processes as neutralization, zero, and distinctive features (morphosemes or semes), the sum of which makes up the invariant of a morphological set.

Since the verb is the most capacious part of speech from the point of view of the number of grammatical categories expressed in it, it would be convenient to illustrate the oppositional method of analysis using a verb of a concrete language, in this case English, which is chosen not by chance but because the English verb, and mainly its tense-aspect system, is a most complex organization without any 'empty cells'.[1]

However, prior to proceeding on to the description of oppositions in the English verb, we should clear up another point, essential for any morphological description, viz., what do we analyse by the method of oppositions: the verb system as a whole, and what is meant by the 'system' in this case, or else should we choose, for a detailed and preliminary analysis, some part of this 'system', and how then will this part correlate with the whole of the system in the plane of oppositions and their *DF*.

In the course of our discussion we have repeatedly used the

[1] On the properties of the verb-systems see Kurylowicz (1953); Khlebnikova (1966a) 69.

term 'microsystem' without properly defining it. Time has come to do it.

In the practice of modern linguistic description there often appear three notions: MICROSYSTEM, MACROSYSTEM and SYSTEM, although each of them separately and, the more so, all of them together have not yet found their explicit interpretation. We shall try to do it using as an illustration the English verb (where necessary).

What is the system of a language? This term (and the notion it contains) is so frequently used, especially during the last two to three decades, that it may seem at first sight that there is nothing here to define. However, the question of how such a notion as 'the system of a language' is related to the distinction of levels of language structure and units of these levels is far from clear. Can we speak of a 'system of phonemes', of 'morphological and syntactic' systems, etc. thus embracing a certain part of the whole system of a language? And how then, say, can the system of phonemes be related to this general system? So, in this part of the discussion concerning the range of applicability of the notion 'system', there arises the problem of isomorphism of the levels of language structure, of correlation of their units, etc.

Furthermore, the term 'structure' is often mentioned when speaking about the system of a language. How do 'structure' and 'system' correlate with each other?

Attempts to specify the notion of 'system' as distinct from 'structure' and to distinguish these notions were made by many linguists; the inconsistency in the content put into these notions and terms, as well as their partial confusion should be noted. Such mixing up will evidently continue, until some formal signs of structure and system will be proposed at the level of mathematical logic. The reason for such a confusion is, however, easily explained. As is known, a structure in any field of science means a complex whole, characterised by a definite set of components, interconnected and interdependent, forming a hierarchy and distinguished by a natural law in their organization, structure, and ties among units. Hence, a structure always implies relations.

The properties of the structure of any object are gradually revealed by science, and the further science advances generally and especially its exact methods, the more clearly are revealed the inner qualities of the object of the given science and the ties among its components, i.e., its structure.

The object of linguistics — language, also shows some qualities of a structure. Language science could not stand aside from the trend to learn the properties of its object — its structure, interconnection of its units, and their relations. From here arose different methods of analysis which were called 'structural', and the notion of structure was introduced into the practice of linguistic research. But simultaneously there appeared a double aspect of this notion. On the one hand it means the material composition of the object, on the other, those relations, which connect language units among themselves. Such a splitting is due mainly to the bilateral nature of those linguistic signs which possess both the plane of expression and the plane of content, each with its own specific structure.

Descriptions of the features of a structure reveal its likeness to a system, since 'system' also implies the presence of natural ties among its elements, as well as a definite symmetry and organization. From this comes the possibility of identifying these two notions and the terms in linguistics.[2]

Common in determining 'structure' and 'system' to many linguists is that the presence of ties among elements of a structure (or system) is thought to be obligatory in both cases. It is with good reason that A. Martinet says that the main hypothesis of structural linguistics is that no part of any object can be understood except in connection with the whole[3].

A purely linguistic definition of structure we find, for example, with P. Hartmann. In the section especially devoted to grammar

[2] For example, Frei (1929) 28-9; Møller (1949) 87-96. Møller expresses an opinion that both terms, which he calls 'the collective term' cover all that is systemic and functional in different units of the language — that is the system of phonemes, grammar and the lexicon.

[3] Martinet (1949) 13-35.

and structure are enumerated all the features of structure as a set of interdependent elements or units linked by structural relations. A structure cannot be an indivisible unity, for it inevitably falls into components; its elements should be analysed separately and together as a unity in their interconnection. Among the elements of a structure some kinds of relations are always present. Only indivisible units are non-structural. P. Hartmann sees the task of a linguist to be the determination of the types of the link among the elements of a structure, their properties, distinctions, etc. P. Hartmann singles out different types of structures *(Struktursorten)*, depending on the nature of their constitutive units (lexical, semantic, grammatical, etc.).[4] It follows from the above, that P. Hartmann also identifies structure with system.

Researchers distinguish various signs of a structure. A structure is a dependence among units, a hierarchy of relations among them, with both the planes of expression and content possessing a structure, says H. Spang-Hanssen.[5] Hierarchy and polarity are two main signs of a structure.[6] Undoubtedly, the admittance of the fact that a structure always implies some relations has had a fruitful influence on the development of methods of linguistic research[7].

[4] Hartmann (1962) 412-417.

[5] Spang-Hanssen (1959) 28.

[6] Jakobson and Halle (1956) 4-5.

[7] In this connection it would be useful to call to mind some definitions of the notion of 'structure' when it is applied to language, and especially when it is discussed by scientists who deal with human thinking and logic, and by linguists of structural persuasion. For instance, Russell (1948) states the following: "To exhibit the structure of an object is to mention its parts and the ways in which they are interrelated" (p. 250). "... an analysis of structure, however complete, does not tell you all that you may wish to know about an object. It tells you only what are the parts of the object and how they are related to each other; it tells you nothing about the relations of the object to objects that are not parts or components of it." (p. 251).

An important concept is that of 'identity of structure'. Russell illustrates it by a linguistic example: "Suppose that in any given sentence you substitute other words ...; then the new sentence has the same structure as the original one." (as in the example: "Plato loved Socrates" = "Brutus killed Caesar") (p. 252). "Every relation has what is called 'a field', which consists of all the

Some linguists still strive to abolish the synonymy of the notions 'structure' and 'system'. A. A. Reformatskij uses the term 'structure' for the vertical cut of the language as a whole, consisting of heterogeneous levels and forming a hierarchy; he calls a 'system' the relations among the signs of the same level.[8] V. A. Zvegincev sees the difference between these two notions in the fact that a structure is a dynamic notion, implying the development of its elements, whereas a system is a static state of interconnected elements.[9]

A structure, in regard to language, is often understood as a linear arrangement of units in a syntagma.

Thus, the representatives of the London school of structuralism consider a structure and its elements as categories extracted from linear combinations in the text, i.e. syntagmatic relations, and a system is considered to represent paradigmatic relations among units which constitute the elements of a structure.[10] "The structure", says M. Halliday, "is a syntagmatic framework of interrelated elements, which are paradigmatically established in the systems of classes and stated as values in the structure".[11]

---

terms that have the relation to something or to which something has the relations". "The relations may be dyadic, triadic, tetradic, pentadic. Dyadic relations are more common for terms of a structure." (p. 254).

The notion of 'structure' and relations as indispensable properties of a structure are discussed in the books of Šaumjan (1962) 85; (1965) 11-14.

Lomtev (1958) 7 is right in affirming that materialistic philosophy departs from the premise that science discovers the substance of things, the relations among which is determined by their objective properties. The goal of any linguistic analysis is the cognition of the objective properties of real linguistic facts, their interrelations in the system of language.

To summarize the above said about 'structure' it is necessary to point out the following: taking into account the properties of 'structure', it may be considered the duty of analysts to investigate the linguistic units which stand in some sort of relation to each other and to identify the KINDS of relations existing among units of different levels of language structure, and among units belonging to one and the same level.

8 Reformatskij (1957) 25-37.

9 Zvegincev (1969) 69.

10 Firth (1957) 1-32.

11 Halliday (1957) 59.

This interdependence is treated in the same way by Ščerbak (1965) 22.

There exists in linguistics also the following differentiation of these notions: a structure is a totality of elements (units) of the given object, while a system is a totality of the ties existing among the elements. The inconvenience of such a differentiation, however, will always be felt when we are to characterize some totalities of language elements, as, e.g. tense-aspect system, case system, etc., which imply both the material composition and meaningful correlations among the components. Evidently, it is difficult to avoid using the notion of structure to designate both the material composition and meaningful interconnections of components of any part of the language which may be designated 'a system'.

It is evident that a strict differentiation between 'structure' and 'system' is hardly possible due to the prevalence of ties of similar order among the components of both of them. It is expedient only to specify the content of these terms and in what way they should conveniently be used in the practice of linguistic description. In regard to the language as a whole, a structure means a totality of its levels, showing definite isomorphism, the units of those levels, in their turn, being organized into systems, the members of which are interdependent and linked by homogeneous relations (e.g. a system of phonemes, grammemes, or parts of speech). If, however, we choose for analysis a certain part of the units of the given level, a microsystem or subsystem (for example, the tense-aspect system), then we have the right to assert that every subsystem, as an object, possesses a complex structure, meaning here both its composition and relations among its components, which unite every system or subsystem into one whole.

In this way the notion of structure in regard to the language as a whole, or to any of its parts chosen for analysis, can be differentiated.

Thus, if a system is a set of signs, joined by the homogeneity of the nature of interconnections among them, then this notion is suitable for denoting both the whole totality of heterogeneous signs, constituting language as a whole, and any totality of units joined by the identity of some common distinctive features (as, e.g. a system of noun declension, of vowel phonemes, etc.), within any level of the language structure.

Whether it is expedient or not in principle to single out different levels of the language, such as the phonemic, morphemic levels, the level of the word, etc., it is difficult to deny a definite, objectively existing succession in the disposition of discrete units of language and speech; thus, one or several phonemes constitute a morpheme, morphemes make up a word, words form various combinations up to a sentence and then a whole utterance. Such a chain reaction, naturally, involves a certain interconnection of units constituting human speech. Establishing the nature of this interconnection is one of the tasks of language science, though it is possible to say beforehand that the very nature of the organization of units of different levels reveals a certain similarity of deep-seated connections allowing the use of equal methods of analysis with only some differentiation conditioned by the nature of the unit under study, when analysing phonemes, grammatical and other word forms, combinations and sentences. It is indisputable, however, that every unit has its own specific features, distinguishing it from the units of other levels.

The isomorphism of various layers of language structure also shows itself in the fact that their units (phonemes, grammatical forms, sentences) exist in two aspects; in other words, they enter into two kinds of links in the language system — associative (paradigmatic) and syntagmatic. In paradigmatics the signs of language are joined into groups (microsystems) on the basis of identity of a definite set of *DF*, in other words, they always form some system; on the other hand, any language unit creates a chain (syntagmatic axis) on which a linear combination of units is taking place according to definite rules of structural organization of the succession of signs.

All these characteristic features testify to the presence of ties, joining heterogeneous units into one common system — language. The same features are a proof that within this many-sided (versatile) totality of elements we may select separate sets of units, joined by the identity of some definite *DF* inherent only for that particular set. At the base of this multilateral structure of language lies a hierarchy of units, which may be characterized as the state in which the units are organized into certain layers which, among themselves, cannot have relations of complete equivalence or complete difference. Units of a language are linked by similar features and relations in which, however, there is always present an element of distinction. Hierarchy is inherent both to the relations among units of different levels and to relations within subsystems of units of the same level (as, e.g., in the paradigmatic system of the verb, in parts of speech, etc.).

Such a hierarchy of relations within the language system leads to a natural desire of the researcher to introduce into linguistic usage the notion of subsystems (minor systems, microsystems, etc.). For example, B. Pottier, subscribing to the definition of a structure as a unity of elements which may be isolated and analysed, considers a system to be the functioning of the elements of the structure. The functions of the elements form a systematics of relations, falling into a plurality of minor systems *(plusieurs petits systèmes)*.[12]

To analyse the verb system of the French language F. Kahn introduces the terms *sous systèmes, sous–sous systèmes*.[13] A. Tellier thinks that the verb is a system of systems and admits the existence of verb subsystems *(sous systèmes)*.[14]

However, in order to have the right to select a small system (microsystem–*MS*) from some general totality of interconnected linguistic signs, there should be proposed definite criteria for the joining of some units into a microsystem, and another notion should also be defined, that of macrosystem. The terms 'macrosystem' and 'microsystem' serve for the contrast of the general and the particular. A MICROsystem enters some MACROsystem, but not vice versa. This does not mean, however, that a macrosystem is necessarily the system of the language as a whole. A microsystem is a further indivisible totality of units joined by the community of some individual, particular *DF* (or features), which serves as the

[12] Pottier (1962) 114, 121 ff.
[13] Kahn (1954) 29-30.
[14] Tellier (1962) 16-19.

basis for distinguishing one microsystem from another *MS* in which a different *DF* appears. Such a set of closely linked units may refer to some unifying system as its component. Then this unifying system refers to the minutest set of units as a macrosystem to a microsystem. The existence of a macrosystem implies the joining of several microsystems. This is the correlation F. Kahn means when he introduces the term *sous système* in regard to *sous–sous systèmes.*

Under any conditions the notions of macrosystem and microsystem are correlative and relative.

Let us illustrate the above with the example of the verb. Thus, (at the morphological level) it would be natural to consider as a macrosystem of the verb the totality of all paradigms of the verb as a class of words liable to inflexion, and as a microsystem, some totality of verb units, joined into one paradigmatic set, in which the units are interconnected on the basis of identity of expressing *DF* (or features) constituting the essence of the given microsystem. Thus, for example, the *MS* of the passive voice differs from other microsystems in the *DF* of passivity, and the microsystem of the subjunctive mood in the feature of unreality, these features being differently marked in the plane of expression. The correlativity of the notions of macrosystem and microsystem may be illustrated by the fact that if we choose for analysis the system of forms of the verbals, then the totality of the paradigms of all the non-finite forms will form a macrosystem, as a unity of non-predicative verb forms, which possess some properties common for all the verbals such as, for instance, absence of person — number, tense (time proper), and mood, and the presence of the marked *DF* of temporal relativity (perfectness). All these features, both unmarked in any exponent, and marked ones serve as a basis for comparison, uniting the verbals into one macrosystem (the past participle, or participle II, stands somewhat apart in the macrosystem, having no perfect and passive sets). The paradigms of each non-finite form would make separate microsystems, in which the basis for unification into one whole (and, simultaneously, of distinguishing them) will be individual *DF*, e.g. categorial and distributional properties of the gerund as distinct from the infinitive, etc.

All similar differentiations and unifications confirm the hierarchical nature of paradigmatic systems and make it necessary to define the principles on which the distinction of microsystems from each other is based in each separate case.

Thus, in the general system of the verb (the MACROsystem) first the contrast of non-finite to finite forms is singled out by the nature of the expression of predication, i.e. by the structure of their paradigms. In their relation to the verb MACROsystem, the finite verb-forms on the one hand, and the non-finite, on the other, will form two MICROsystems, further indivisible by the nature of predication; however, they also are macrosystems in themselves which can be further divided, if we take as the basis for comparison other features — not the way of expressing predication, but more particular categorial markers characteristic for form-sets of each verbal.

The nature of correlation between a macrosystem and microsystems will also be varied depending on the nature of the units joined into one set. Thus, the macrosystem of verbals consists of microsystems, subdivided according to the type of the non-finite form, that is, the microsystem of the gerund, the homonymous microsystem of participle I (which forms a special microsystem different from that of the gerund by the nature of the categorial content and valency of its units), or the microsystem of the infinitive. As has been mentioned above, participle II falls out of the typical structure of paradigmatic sets usual for other verbals. Perhaps it can be considered as part of the participle *MS* with a latent *DF* of passivity. As our main task here is the oppositional analysis of a finite *MS*, we shall not go deeply into the problem.

These microsystems differ from each other in a number of essential features — formal, distributional, and semantic — and the elements in them are joined together on the basis of some common features such as classes of units similar to parts of speech. Hence, the nature of contrast within the non-finite microsystems is different as compared with the finite ones. The contrast of verbals to the finite forms is built not only on the *DF* of abstract verbal categories, natural for the action expressed in the verb, such as tense, aspect, and voice, but on a syntactic basis depending on the ability to include the person of the action, this ability directly reflected in the nature of valency of the non-finite forms in speech compared with the finite ones.

Within the finite forms, units are joined into microsystems not as different classes of units but on another basis. This basis is COMPATIBILITY–INCOMPATIBILITY OF DF (semes) as the minutest units of the plane of content (grammatical category) within one microsystem and in one categorial form.

Thus, passivity and activity are incompatible *DF* contrasted to each other. They show differently the relation of the action to the subject and, correspondingly, influence the correlation of the main members of the sentence, changing the sense of predication. Hence, passivity and activity, as semes of the action of the verb, cannot be expressed in one microsystem or in one categorial form. Passivity, as a marked *DF*, has a special exponent; in English, for instance, it is expressed in the combination to *be* with participle II. This way, the minimal difference between two microsystems will be one *DF*, present in the given microsystem and absent in another one, contrasted to it as an unmarked member. Thus, in the active tense-aspect *MS* this feature is absent. This relation, both in the plane of expression and the plane of content, is clearly seen in the comparison of two categorial forms, e.g. *has done* and *has been done*, where the addition of one *DF* is exposed in the structure of the form.

On the other hand, tense, aspect, and perfectness (temporal relativity) are compatible semes, intersecting in one categorial form and, correspondingly, in one microsystem. In the word form *has been doing*, semes of perfectness, present tense, and duration are combined. These semes, being signs of grammatical categories of the verb, are compatible with any marked *DF*, constituting a special microsystem, viz., with semes of passivity and unreality.

That is why the microsystem of tense-aspect, active indicative, containing the forms which express compatible semes of tense, aspect (duration), and perfectness, is the basic, initial microsystem; when other *DF* are added, new microsystems are formed. It is the most capacious and differentiated *MS* containing the greatest amount of members, the categories it expresses being most differentiated precisely when the subject is active and the action real (and not passive or unreal).

The seme of unreality finds formal expression in unreal moods (or mood — it depends on the language) and, hence, forms a special paradigm or microsystem. However, in English, in the *MS* of unreal moods the expression of basic verb categories, especially that of tense, has a specific nature of its own. Taking into account certain distributive factors which suspend homonymy, this mood is represented by a special paradigmatic set, incompatible with the paradigm of the tense-aspect microsystem.[15]

Thus, on the basis of compatibility or incompatibility of various grammatical *DF* in the system of finite forms of the English verb, three microsystems become clearly defined, having different structure of the formal means in different combinations. There exists precisely three and not four microsystems, since one of them, the initial one (of tense-aspect) is a carrier of both active and real actions, and the other two are marking unreality and passivity, i.e. in each of them one relvant *DF* is added.

The above shows that the relations within the MACROsystem of the verb may be both intraparadigmatic (among members of the same microsystem) and interparadigmatic (among separate microsystems).

Traditional grammar has long since implicitly proceeded from the varied combinability of grammatical *DF* grouping into special paradigms the conjugation of the verb in the active, passive, and subjunctive mood, since, as a rule, it considered the actual state of the distribution of the material means of the language. This inductive distribution of verb forms among separate paradigms is theoretically substantiated on the basis of *DF* and their oppositions within and beyond the limits of every separate microsystem.

15 On the problem see Khlebnikova (1965b).

The problem of compatibility of heterogeneous or homogeneous features is of great interest for investigating the inner structure of different units of a language, the compatibility to be considered both in the plane of expression[16] and in the plane of content of bilateral units of language, as well as in regard to phonological characteristics.[17] Compatibility or incompatibility of formants or *DF* may be regarded both in associative (paradigmatic) and linear (syntagmatic) planes, and this question should be the subject of special study.

Interparadigmatic oppositions differ from intraparadigmatic ones in that they include a contrast of the maximally generalized grammatical feature, which is formally marked in one of the microsystems. Thus, passivity is marked in one paradigm only, as is unreality, while in another paradigm (microsystem), unmarked in the plane of content by a voice or modal *DF*, these *DF* are not marked at all in the plane of expression. A specific feature of interparadigmatic oppositions is also a large number and volume of constituents of each of the two contrasted members — of the microsystems.

These three microsystems are interconnected as members of privative oppositions, presenting two oppositions — in mood and voice, with one common unmarked member — the tense-aspect *MS*. In microsystems, strong, marked members of interparadigmatic oppositions, there is always present one categorial *DF* (passivity, unreality) which is absent in the weak member of these oppositions — the *MS* of tense-aspect active. On the other hand, the distinctive features present in the initial member of interparadigmatic oppositions (tense, aspect, perfectness) can potentially enter other microsystems and serve as a basis for comparison of interparadigmatic privative oppositions.

The general interrelation among finite microsystems, and their hierarchy may be presented as a scheme of concentric figures

[16] On co-occurrence and incompatibility of verb exponents see W. Twaddell (1960) 1-4. Formal incompatibility must undoubtedly be conditioned by incompatibility of *DF* in the plane of content.

[17] Cf. Revzin (1964) 59-65.

which shows that the tense-aspect, active *MS* is the centre and privative member of interparadigmatic oppositions among microsystems, since marked strong *DF*, constituting its basis, viz., the *DF* of tense, duration and perfectness, are potentially capable of entering other *MS* as well (those of passive and unreal mood), but not vice versa. These *DF* serve as a basis for comparison of countermembers of the two oppositions in voice and mood.

In the passive *MS* one *DF* is added, that of passivity, so that four strong *DF* are present (marked) in it, all being reflected in formants, e.g., *has been done, was being done.*

These four *DF* can enter a third *MS*, unreal mood, in which another *DF*, unreality, is added. Unreal moods, however, always present a mixed picture in different languages in the plane of markedness of the *DF* of unreality and other *DF*. For example, in English, this *MS* is devoid of the *DF* of tense, while the *DF* of unreality is realized in it only on the syntagmatic axis, under certain conditions of environment due to the all-round homonymy of formants of this *MS* to the other two *MS*, for example, *if it had been known before, everything would have seemed different.*

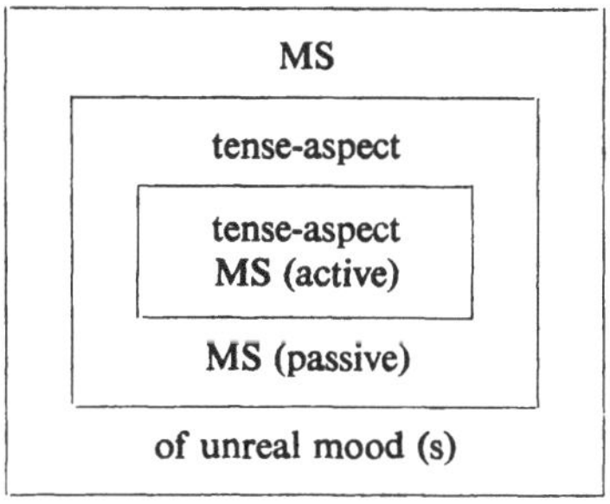

Thus, microsystems are singled out on the basis of strong, marked *DF*. This fully corresponds to the principles of functional paradigmatic analysis. If, however, we turn to the habitual canons of the normative school of grammar, we shall see that paradigms of active and passive voices, indicative and unreal moods, will be found in different *MS*. This is quite natural and logical, if we consider the plane of expression and the plane of content proceeding from the essence of the *DF* constituting them in the given language and from the presence of exponents of these *DF*.

Interparadigmatic oppositions may be also presented in the following way:

MS of tense-aspect active
(unmarked member)

← ——————————→ O ←——————————→

MS of passive (marked member) | *MS* of subjunctive mood (marked member)

The lines show the direction of two privative oppositions with the privative member, common to both of them, represented by the tense-aspect *MS*.

Interparadigmatic oppositions are bilateral, since the *DF* expressed in their strong member is not repeated in other microsystems. As to the intraparadigmatic oppositions, they are all multilateral because they are proportional, the features of one opposition being repeated in the other, acting in the next parallel set. Thus, e.g. the correlation of perfect — non-perfect sets or durative — non-durative in the tense-aspect *MS* is similar on all temporal axes.

As was pointed out above, the unmarked member of a privative opposition possesses a more general meaning compared with the marked one. Hence, in the case of interparadigmatic oppositions containing numerous co-members, the *MS* unmarked by the *DF* of voice and mood is the basic, most widely used *MS* and penetrates into all the other microsystems, because temporal and aspect categories are potentially possible in the other two microsystems. There is not a single form in the microsystems of voice and mood, that would not draw its formal expression from the scheme of tense-aspect as the central microsystem, determining the composition and categorial content of forms in other paradigms. Hence, an effective analysis of one marked microsystem is impossible without a prior study of the basic microsystem of tense-aspect; for this reason that *MS* is the main object of our description. An especially close interconnection may be expected between tense and mood, since the mood is but transporting the action into another modal plane without changing the nature of predication and subject-object relations of the verb, thus allowing the microsystem of mood to have both active and passive forms.

The correlation of microsystems among themselves shows their hierarchy and the absence of identity among them. This is the main condition of the existence of a morphological system among the members of which there are no equivalent relations, but a distinction built on a definite identity, which serves as the basis for joining heterogeneous units into one macro- or microsystem.

If we consider the verb as a macrosystem, consisting of microsystems, and if we characterize microsystems by the quality and number of grammatical *DF* (semes) they express, then, naturally, a question arises, in what way the MACROsystem itself can be characterized, and how its *DF* correlate with the *DF* reflected in microsystems. A macrosystem as a whole may be presented as a set of semes and their formal exponents which are distributed among microsystems, depending on the nature of markedness-unmarkedness of certain *DF* in corresponding microsystems. The following table shows that the macrosystem of the English verb contains a definite set of semes having their own exponents:

semes *(plane of content)*[18]	**exponents** *(plane of expression)*[19]
(1) tense (depending on the individual *DF* of the axis of orientation)	(1) base (2) base +– ed (3) shall, will / should, would } + base
(2) duration (aspect)	be + base +– ing
(3) perfectness	have + base +– ed
(4) passivity	be + base +– ed
(5) unreality	exponents homonymous to the past or future II

[18] The *DF* of person–number are not reflected in this table because they characterize different members within one and the same paradigm, and they do not enter any oppositions outside that particular paradigm.
[19] Irregular formations are not taken into account here. The standard pattern of verb forms is taken as basic for the English verb.

How this set of *DF* is distributed through microsystems, and how *DF* and their markers combine in one microsystem and in separate wordforms is cleared up in the process of analysing every microsystem. The interrelation of *DF* and their exponents is revealed on the basis of oppositions and, since in any opposition a hierarchy of its members is necessary, the notion of markedness-unmarkedness is introduced into the correlation. In the plane of expression in the macrosystem, synthetic forms are contrasted to analytical ones, the former entering as a component into all the analytical forms. The constant intersection of grammatical *DF* and their exponents in separate microsystems is the essence of the whole verb system. The fact that microsystems exist in the verb is an illustration of the principle of modern linguistics that language is a system of systems.[20]

Morphological microsystems consist of units, carriers of a certain grammatical content. The form and content constantly interact, and the study of the nature of this interaction must also be the subject of morphological research. The plane of expression, as a directly observable fact, can be studied by observing the order and structure of morphemes, while the content is not directly observable and thus, its analysis calls for abstract thinking. These aims are served precisely by oppositions, which explain with maximum simplicity the correlation of the units of every micro- and macrosystem.

Thus, the three notions considered above are interdependent among themselves, and the manner of their interdependence shows the hierarchical nature of language units and their combinations.

[20] Cf. Daneš and Vachek (1964) 21-31.

10.

OPPOSITIONS IN THE TENSE-ASPECT MICROSYSTEM OF THE ENGLISH VERB (THE APPLICATION OF THE THEORY OF OPPOSITIONS)

Having defined the notion of microsystem, we can now begin an oppositional analysis of concrete material, viz., the tense-aspect *MS* of the English verb. Let us take the following principles as initial: in microsystems of the verb only such formal units can exist as possess the following properties: they must be universal and cover the whole bulk of the verbal lexicon (with a very few exceptions) in accordance with the meaning of the category they express; they must be recurrent in parallel rows and series. Redundant forms are also subject to these principles. Paradigmatic units must be formally homogeneous, the more so if they are homonymous. Such are the general rules of the structural organization of morphological sets which express grammatical categories. Chance elements cannot be considered as belonging to a paradigmatic set, because a grammatical category is characterized by maximum abstraction from individual meanings, and, hence, the forms expressing it cannot deviate from the standard established in the given language. Thus, e.g., such combinations as *got tired*, *became bound* cannot be ascribed to the means of expressing the passive (contrary to the assertion of some grammarians),[1] since they do not possess universality and reiteration in all the sets, possible for the given microsystem in the given language. The same is true of combinations like *was gone*, *am gone*.

Such units present a special phenomenon of morphological

[1] For example Curme (1935) 219 ff.; Roberts (1954) 129.

variation (not considered here),[2] and they cannot be numbered among the exponents of some grammatical category, due to strict limitations in their form-building potential.

According to the tasks set here, we shall consider only the invariants of paradigmatic sets of the tense-aspect *MS*, their *DF*, the nature of oppositions formed by these sets, and their neutralization.

Since the composition and correlation of the members of the given microsystem are a point of controversy, we shall dwell in short on some essential deviations from the scheme established here and their causes. Though a detailed syntagmatic analysis is not a part of our task, some examples will be given to illustrate the invariants detected; also on the basis of the proposed method of analysis some recommendations for syntagmatic analysis of concrete language material will be given, and some illustrations will be presented of the actualization and neutralization of oppositions in speech.

Since the tense-aspect *MS* is the initial and basic one in the verb system, those verb categories are marked in it which are essential for the whole verb system and may find expression in other microsystems. These are the categories of tense, temporal relativity (perfectness), and aspect (durative), the most universal of all the other verbal categories, capable of entering other microsystems.

Being an unmarked member of the opposition in voice and mood, the tense-aspect *MS* should be called the *MS* of tense-aspect, active voice, indicative mood. That part of the name, however, which implies its functioning as a member of privative INTERparadigmatic oppositions, is suspended, since, for analysing oppositions WITHIN a microsystem, its quality as a member of interparadigmatic oppositions is insignificant, and in its grammemes, those *DF* which are contained there as unmarked *DF* of interparadigmatic oppositions (active voice and indicative mood), are irrelevant.

[2] The problem of variation within the paradigmatics of morphological units is dealt with in Khlebnikova (1970).

We define this microsystem with a term in which the *DF* of tense (and not aspect), comes first, since it is tense that forms the frame of the system (and not aspect), though opinions differ on this score.

Thus, J. Kurylowicz thinks that the constituting content of verb forms is aspect, and not tense, which is only attached to the aspect, like the article is attached to the symbolic content of a noun. The notion of the moment of time, according to J. Kurylowicz, does not create a content of qualitative order but is simply an indicative element.[3] Surely it is true that aspect is a category which serves to indicate the qualitative characteristics of the action; it does not mean, however, that in all languages aspect must be the basic category, dominating all the other verbal categories. Depending on the trend of the development of the verb system in different languages, the basis on which the finite verb paradigm is built may be expressed either by an aspect or tense division. And, although tense, as distinct from aspect, possesses an indicative character rather than qualitative, it is for the verb as a part of a speech denoting process that the temporal characteristics of the action are most essential.

Tense, as an inherent property of the verb action, is a complex event. It is not without reason that linguists tried to create some typological scheme of tense valid for all languages, which would reflect all possible temporal divisions that may find formal expression in the morphology of the verb in each language.[4] Such logical schemes (constructs) created by grammarians (and sometimes by logicians) are not universal with all linguists either, and bear the traces of the researcher's point of view and of the language material upon which they are based. As is known, O. Jespersen proceeds from seven notional tenses, considering the present to be the centre of notional time from which a line goes left, into the past, and right, into the future, both past and future being capable of having three divisions, indicating simultaneity, anteriority, or posteriority.[5]

[3] Kurylowicz (1953) 109-18.
[4] Cf. Kurylowicz (1956) 26.
[5] Jespersen (1951) 257. Cf. also Diver (1963) 141-181.

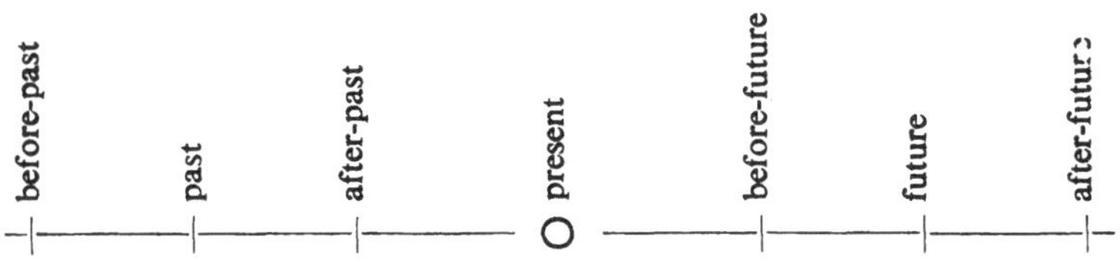

It is noteworthy that no language contains a grammatical form which would express the posterior past or posterior future. Different authors cite various artificial examples which include combinations not entering into the system of grammatical forms. Thus, H. Garey gives the example *I'm leaving in a minute* as posterior present, while he thinks posterior future to be but a theoretical possibility, illustrating it with the latin *amaturus ero.*[6] O. Jespersen gives as an example of anterior past such combinations as: *Next year she gave birth to a son who was to cause her great anxiety*, adding that he knows of no language in which this stage would be expressed in a simple form.[7] This is not accidental. The question is what temporal landmarks can be expressed by the verb and upon what may they be based.

Since every grammatical category is indirectly linked with categories of thought, the most exact approximation to the real, notional time will be the division into present, past and future, according to the manner in which the events expressed by the language material are considered. It should be noted, also, that the main divisions are those of present and past as directly observable points, and in regard to which exact information can be produced supported by facts. That is why A. Isačenko defines the past as expressing the process which can be viewed, and the present as processes, taking place at the moment.[8] The future, as a projected tense, cannot but contain an element of supposition, i.e. modality; hence, some languages do without reflecting this tense in verb forms at all, others use modal formants to denote it, still others make use of special form-building exponents, etc.

6 Garey (1955) 13-14.
7 Jespersen (1951) 262.
8 Isačenko (1960) 420-25.

The future seems always to stem from the present, though it is in opposition to it, being the stronger marked member of the equipollent opposition. The essence of a future action, as the one not directly observable, is reflected both in its form in different languages (in its secondary formation on the basis of modal or close to modal formants), and in its treatment by linguists. As is known, far from all grammarians recognize the existence of future as an independent verb form of temporal order. Thus, J. Kurylowicz calls the future a mood sometimes denoted by clumsy terms like subjunctive and conjunctive, due to the fact that it expresses possibility, probability, and expectation. However, J. Kurylowicz contrasts the future to the present + past, i.e. to categories of tense, which is contradictory.[9] B. Strang is inclined to consider the pure future as a rare category, identifying the formants of future *shall* and *will* by their formal distributional characteristics with modal verbs, suspending, in fact, the problem of their grammatization.[10] Due to the modal content of the future and the secondary nature of its formants, often any means with a shadow of futurity, up to adverbial clauses of purpose, are referred to as future. In A. Tellier's book, the system of tenses also includes combinations with verbs *has to*, *was to*, etc.[11] A. Tellier considers the future to be a mood, but says that the future has originated from the present, whereas the conditional has sprung from the past. It is clear that in this definition notions of different planes, temporal and modal, are entangled.

In any case, the three temporal divisions with which the events of reality can correlate, somehow or other find reflection in the grammatical category of tense and are the basic axes of orientation for the verb action.

However, the complexity of the processes of human thinking and their reflection in the language leads to further temporal differentiation within the axes of time-orientation. Most European languages do not limit themselves to the above three divisions

9 Kurylowicz (1953) 109-118.
10 Strang (1968) 167-168.
11 Tellier (1962) 327.

to localize the events in time. Potentially possible are various RELATIVE temporal points, within the framework of which events take place, but these points are necessarily linked with the three basic temporal divisions. In this way the system of relative tenses develops, the frame for which are the three basic divisions of time. Relative tenses form additional landmarks on the axis of each real tense. These relative points on the basic axes of orientation are able to denote only two stages of action in regard to the axis of orientation: an action simultaneous with the time denoted by the axis, and an action preceding it. No posteriority can be denoted by a relative temporal division, since it is a purely temporal notion and not a relative one, and it is realized on the scale of time and not within the limit of every axis of orientation. That is why on the scheme of O. Jespersen we find no agreement between the point of posteriority and verb sets existing in the languages.

A hypothetical system of time was built by W. Bull, who took into account the retrospective–prospective axis of orientation which localizes the action as future from the point of view of the past.[12] This is not so much a notional, extralinguistic scheme, as a scheme at the level of universal grammar, generalizing all those possible temporal meanings which can be expressed by the verb in Indo-European languages. The axis of orientation, which includes narrated events proceeding from the past, found its grammatical realization, in particular, in French and English in a special series of forms of relative future ('future in the past'). The fact that such an action creates a special temporal axis of orientation is proof of our supposition that posteriority is a purely temporal category, and not a relative one; for this reason, *F*II is potentially capable of having verb paradigms, inherent for the three basic axes of orientation, i.e. such sets as continuous and perfect.

The scheme of temporal divisions in H. Reichenbach's work on symbolic logic is built on the basis of logical relations.[13]

[12] Bull (1960) 26. Cf. the psycholinguistic schemes in the books Guillaume (1929) 34; Brunot (1953) 438 ff.
[13] H. Reichenbach (1947) 288 ff.

This scheme does reflect those divisions which can denote verb actions by means of morphological forms of the English verb. Here, however, as with O. Jespersen, the existence of the stages of posterior past and posterior future is implied. These stages, however, find no equivalent in any existing grammatical form, though H. Reichenbach states that the form *would do* has the meaning of posterior past, and he looks for posterior future in modal verbs.[14] H. Reichenbach defines the temporal relation of the English verb forms as proceeding from three facts: point of event — E, point of reference — R, and point of speech — S. This procedure seems to us most useful for localizing the event in time. Thus, for example, considering any event on the basis of its relation to the above three points, all the English verb forms are placed on the horizontal line, including the three points *ERS* in different relation to each other. For example, the event *I had seen John* is represented in the following way: $\overset{E}{\circ}—\overset{R}{\circ}—\overset{S}{\circ}$. It means that the event takes place prior to the point of reference, precedes it, while the point of reference in its turn precedes the moment of speech. In the event *I see John* all the three points coincide, etc. Continuous forms are denoted as forms showing the duration of the event within the same point of reference, as, e.g. *I was seeing John* is presented as $\underset{R,\,E}{\boxed{\quad\quad}}\ \underset{S}{\circ}$. H. Reichenbach also takes into account the point *F* II (p. 293), for the very reason that he uses logic in his treatment of the content of temporal forms.

Thus, essential for specifying the essence of the verbal grammatical categories is the direction in which the action is observed in regard to the temporal axes, as well as the intervals between the actions. To consider the meanings and functions of verb forms from these points of view, W. Bull uses vector (direction finding) and scalar tensor formulas (defining the nature of intervals between the actions).[15]

The attempt by W. Bull and H. Reichenbach to place the forms of the conditional mood on the scale of time (tense) can scarcely

14 Reichenbach (1947) 297.
15 Bull (1960) 14 ff.

be considered successful, since its modal content brings it beyond the sphere of exact temporal divisions, and its formal data, in English for example, deprive it of the ability to include an event, independently of the context, into a definite temporal framework. But, on the other hand, the consideration of the logical motivation of temporal divisions in the language helps both W. Bull and H. Reichenbach to find the place for such important notions as simultaneity and anteriority and include them into the scheme of temporal divisions.

What, then, is the basis of the English scheme of tense-aspect? This paradigm is based upon temporal divisions forming a frame in which aspect differentiation is included within the range of different temporal points. The English language, as J. Kurylowicz rightly points out, possesses a verb system approximating a complete one;[16] this means that it contains grammatical expression not only of temporal relations, more or less corresponding to the real time (present, past, and future), but also of relative temporal meaning of simultaneity and anteriority, realized within the axes of orientation which coincide with the actual temporal divisions. The completeness of the English system, which has no empty cells, is also expressed in the reflection of the retrospective-prospective axis of orientation, which does not correspond to any division of real time, but is a purely grammatical manifestation of temporal relations, caused by the distribution of verb forms, the above distribution creating a situation of temporal dependence of a future projected action upon the event in the past, when the narration of future events is taking place within the temporal point of the past. However, this dependent axis of orientation possesses the same qualities, in its formal filling, as the three independent axes of orientation, which logically correspond to those temporal divisions which the human brain comprehends as independently existing or accessible for observation.

To distinguish time as an extra-linguistic and a linguistic category, the English language has the different terms 'time' and

[16] Kurylowicz (1953) 109-18.

'tense'. Temporal denotators, i.e. forms denoting the grammatical category of time, may be defined not by their relation to the signified (the referent of reality), but exclusively depending on the system of verb tenses in the given language, on their relation to each other. Both the meaning of grammatical forms and their distribution in speech must be defined only in terms of a concrete system.

Proceeding from the above, such a notion as the moment (act) of speech should be considered as a grammatical one, since it serves as the starting point of temporal relations in the verb system, and in this connection A. Isačenko rightly notes that it does not have a correlate in empirical time due to it being an infinitely short period.[17]

In oppositions of the verb system of the English language the present, or the moment of speech, also serves as the initial member of a set of oppositions.

To conveniently observe the microsystem of time-aspect, we shall produce here this scheme in its written form (page 112).

This scheme preserves traditional terminology which it is not sensible to change; however, for the sake of brevity and convenience the terms Future II (as well as Future I) are introduced, since the figures I and II attached to a common term generally mean that two units under discussion are linked between themselves by the community of their categorial meaning and by their belonging to the same class of units (compare the denomination of participles I and II).[18] Thus, the future tense in English is split into two independent axes of orientation, one of which stems from and depends on the present, and the other on the past. This temporal dichotomy may be illustrated in the following way:

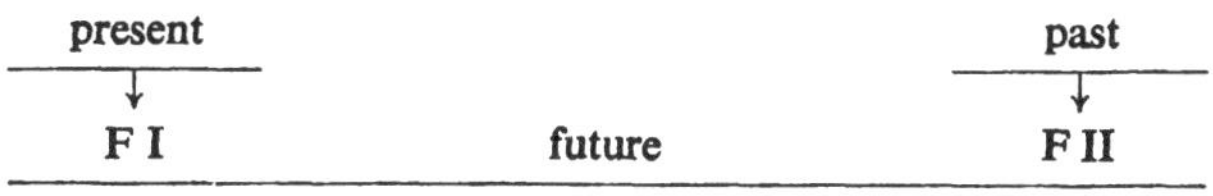

[17] Isačenko (1960) 420.

[18] The traditional terms 'Present Participle' and 'Past Participle' do not reflect correctly the place which is occupied in the grammatical meaning of these forms by the *DF* of tense. In reality the *DF* of tense is not expressed in any way in the forms themselves.

The paradigmatic scheme of finite verb-forms of the tense-aspect microsystem (active indicative) *

	Non-Perfect		Perfect	
	a) Indefinite (I)	b) Continuous (C)	c) Perfect (P)	d) Perfect Continuous (PC)
I Present (Pr.)	1st pers. sing. move	am moving	have moved	have been moving
	3rd pers. sing. moves	is moving	has moved	has been moving
	pl. move	are moving	have moved	have been moving
II Past (Pt.)	sing. moved	was moving	had moved	had been moving
	pl. moved	were moving	had moved	had been moving
III Future I (F I)	shall, will move	shall, will be moving	shall, will have moved	shall, will have been moving
IV Future II (F II)	should, would move	should, would be moving	should, would have moved	should, would have been moving

* Forms with the auxiliary verb *do* fail to express any special grammatical category of the verb, hence, independently they do not create any oppositions, being purely syntactically defined; therefore, they are not considered here. Since the grammemes with shall (should) and will (would) as auxiliaries are no longer strictly limited to a certain person, the person-number differentiation is not reflected in the table.

Each axis of orientation is divided into two parts, the first including non-perfect forms, the second perfect ones. This temporal division on the horizontal axis corresponds to the reflection in verb forms of relative temporal divisions, situated on each axis of orientation. It is the presence of forms which imply anteriority that favors the rise in the language of a system of sequence of tenses on the syntagmatic level. Those languages which do not have any special forms to express anteriority (like Russian) are devoid of such a formal sequence of tenses. The system of forms, implying either simultaneity or anteriority to the axes of orientation, has a different categorial content compared with the grammatical tense proper. Different authors, including H. Reichenbach (in the book cited above), deal with relations to the point of reference and consider the perfect to be a relative tense, though few consider perfectness as a grammatical category. Thus, for example, W. Francis calls the correlation of perfect–non-perfect two phases.[19] R. Jakobson makes use of the term taxis for relative tense.[20]

Unreservedly joining the linguists who accept the relative nature of the perfect, we call the opposition of non-perfect–perfect a category of temporal relativity (perfectness), expressing reference to the temporal axis of orientation.

Perfectness is the most enduring and essential category, acting in all microsystems of the English verb, and its categorial essence in the microsystem of tense-aspect presents a special interest due to the specific nature of the structure of the axes of orientation, which represent a grammatical reflection of temporal distinctions existing in life.

The basic condition for the examination of the structure of a tense-aspect system is a strict differentiation of two different temporal categories — 'tense' proper and 'perfectness' ('temporal relativity'). The recognition of perfectness as a quality of each axis of orientation separately, as a category, active within the limits of every axis without overstepping it onto another axis of orientation, implies the recognition of it on the axis of the present

[19] Francis (1958) 330. See also Joos (1964) 138.
[20] Jakobson (1957) 4.

as well. If, however, we return to the notional scheme of O. Jespersen, we shall see that it does not hold the perfect set of the present *(Pr.P.)*, since anteriority to the present is marked on it as the past which, however, is a special temporal axis of orientation and has nothing to do with the relations of simultaneity and priority to the axis of the present. Anteriority to the present is not at all equal to the past, since the past is a division of TIME, constituting an independent axis of orientation, while anteriority (reflected in perfectness) is the nature of reference of the action to the plane of the given axis of orientation, no matter what are its purely temporal characteristics, whether it is the axis of the future or past. The very basis of the scheme of temporal divisions makes O. Jespersen rule out the perfect and consider it outside of tense and aspect.[21]

The root of the argument in treating the essence of paradigmatic verb forms in different languages is precisely the consideration of the verb tense without dividing it into tense proper, marking the temporal axes of orientation, and the relative tense (perfectness), showing the intervals of the action, the relation to the point of temporal reference *(Beziehungspunkt)*,[22] and to the axis of orientation in time. Temporal relativity cuts vertically through all the axes.

Here is a schematic representation of this correlation:

	anteriority to the axes	
axis of the		present
axis of the		past
axis of the		future I
axis of the		future II

One of the main causes of disagreement on the question of ASPECT is the lack also of such double differentiation of temporal categories, since in the languages where the base of the verb

[21] Jespersen (1951) ch. XX.
[22] Weber (1954) 25.

paradigm is made up of temporal and not aspectual relations, the aspect is distinguished within the framework of temporal divisions, and not vice versa. The treatment of aspect is also influenced by the aspect scheme of the Slavic languages, which is based on the distinction between the perfective and imperfective aspects, and which has subordinated the distinction both in tense and, mainly, in temporal relativity. Temporal relativity is absent where the base of the verb paradigm is an aspectual distinction of perfective — imperfective and, hence, the formal sequence of tenses is absent too. Aspect in this case seems to absorb the differentiation of simultaneity and anteriority and, on the contrary, where perfectness enters the system the opposition of perfective — imperfective aspects is absent.

Simultaneity as a temporal category always implies non-completeness and duration, while anteriority implies completeness; thus we observe an especially close intertwining of aspectual and temporal shades in these two planes; here is manifest the mutual dependence of aspect (as a category implying the nature and the character of the action) and perfectness (as a category marking the nature of the reference of the action to the temporal axis of orientation). It would be appropriate to cite here A. Isačenko's remark that in the notion of completeness the boundary between temporal and purely aspectual semantic elements disappears.[23] The prevalence of temporal or aspectual meaning again depends upon the established system of the verb, on whether it has taken the path of the domination of aspectual or temporal relations.

Such intertwining of aspectual and temporal meanings leads different linguists to the treatment of perfect and non-perfect sets as either aspectual or temporal. However, a tendency should be noted of late towards a more logical understanding of the essence of relations in the verb paradigm and, hence, towards understanding the relative nature of perfectness.

Such a trend is manifest first of all with representatives of logic (H. Reichenbach) and also with linguists; J. Aiken, H. Weber,

[23] Isačenko (1960) 134.

W. Bull, A. Markwardt, F. Kahn, H. Garey and others, as well as in the Moscow school with A. I. Smirnitskij and his followers in this question.

We have established so far that the frame of relations within the English tense-aspect MS are relations created by the grammatical category of tense expressed in contrasts of the axes of orientation, and by the grammatical category of temporal relativity (perfectness) marking two interdependent, relative temporal divisions, those of simultaneity (unmarked member) and anteriority (marked member) on each axis of orientation; it is only natural that temporal relativity acting only on the given axis of orientation signifies close contact with the temporal axis. Aspect (duration) differentiation is taking place exclusively within the temporal relations of both types, constituting the frame of the whole microsystem. By means of tense we understand the correlation of an event with reality in the plane of its localization in a certain time, while through temporal relativity we learn the nature of intervals between actions in regard to the given axis of orientation. "Tense is a set of responses to temporal conditions",[24] since all events take place in time, the act of speech always being the initial point of orientation for a verb system. This is an important principle to be considered when analysing any kinds of relations in the verb.

In analysing relations in the verb system, it would be natural to consider not only the function and meaning of every verb form separately but, above all, its systemic qualities, manifesting themselves in its relations within the given *MS* and in the macrosystem as a whole. That is why the meaning and essence of forms may best be understood through oppositions, which reveal the qualities of a grammeme as part of the whole, and show the invariant which is contained within every form as a point of intersection of relations in the system.

The unit of verb conjugation analysed by means of oppositions is denoted here by the term 'set' (or paradigm), by which we shall

[24] Kantor (1936) 247.

understand a group of forms of verb conjugation joined together by common grammatical features and differing in person and number.[25] In other words, the specific nature of a set (which always contains forms of the same type or subclass) is defined by the relation of the given individual paradigm to the leading verb categories, abstract in their essence and characteristic precisely for the verb as a class of words denoting action or process (while both person and number may appear in other word classes and do not belong to the verb only). Thus, the heterogeneity of verb categories influences the content of the set and its ability to respond to the relations prevalent in the verb. Only its markedness in purely verbal, abstract grammatical categories is relevant for the place of the set in the scheme of relations. For instance, a set is a paradigm of the conjugation of the forms of the present continuous, or past perfect, etc. However, for greater convenience in treating the material in the process of analysis, it would be expedient to call also 'a set' a whole series of forms of the sphere of present, perfect, etc. Then, for one unit of verb conjugation we can use the term 'paradigm'.

In analysing the verb, such a term as 'formtype', is also used when a definite morphological model of verb formations is meant, and when it serves as an exponent of some grammatical category (in some cases the notion of 'formtype' may coincide with the 'set'); yet another term is applied — 'wordform' — by which a separate categorial form is meant, e.g. third person singular.[26] The generalizing term for all possible verb formations is the usual term 'form', suitable for practical use. The term 'grammeme', defined by us for denoting a verb form as a cluster of *DF*, also remains valid.

[25] Cf. Pospelov (1953) 206-246.
[26] Smirnickij (1957) 30.

11.

OPPOSITIONS IN THE CATEGORY OF TENSE

Our first task will be to establish the nature of time relations in the scheme of tense-aspect, all the paradigms of which are joined into one whole.

Defined as one of the principles of the functioning of the verb system, the principle of the binarity of contrasts remains the basic one in the correlation of those sets or formtypes which represent the essence of the grammatical category of tense. The prime point of all the verbal relations in the plane of time is the point of the present moment (the act of speech), since man's account of some event is a real and objective criterion of the relations existing in extra-linguistic reality, interpreted through the prism of human brains. The moment of speech is the main axis of orientation (temporal plane) of verbal actions, which gives rise to other temporal planes. Priority to this moment is expressed in the past, which is not linked with the moment of speech and creates a second axis of orientation — the axis of the past. Anticipated events posterior to the act of speech find expression in the following axis of orientation, marked by the term 'future'. But the succession of events may be considered not only as proceeding from the axis of the present, but of the past as well. Herein lies the specific nature of the interpretation of events through the prism of man's attitude to the time of speech. Hence, there can exist (as, e.g., in English or French) another axis, that of a future action considered from the point of view of the past (*F* II). All these four axes are represented by verbal forms in the English tense-aspect *MS*, so that they cannot be considered an extralinguistic category; they form the essence of grammatical events and relations as a grammatical category, on the line of tense expressed, first of all, in the indefinite set of forms, not complicated by the *DF* of duration or perfectness; these four temporal axes are the core of the realization of different relations characteristic of other verbal categories (cf. the correlation of *has been writing* and *had been writing*, where the grammatical difference between the forms is reduced to the distinction of the *DF* of tense — present or past).

Axes of orientation, like any other categorial units, express relation, in this case a three-plane relation, that of priority, simultaneity, and succession (posteriority), all of which are regarded from the moment of speech. This relation is interpreted in a series of binary oppositions in paradigmatic sets. The grammatical nature of the temporal axes is confirmed by the fact that not all the axes are found in some languages (e.g. in Russian there is no axis of succession from the point of view of the past), as well as by their markedness in certain morphemes (including zero ones) of indefinite forms.

Each axis includes not only separate grammatical forms, denoting tense, but, according to its term, it is an axis of other categorial relations, and in particular, of relative temporal relations expressed in perfect–non-perfect sets. This opposition is relative because its members interact only within every separate axis, not overstepping its boundaries. Hence, the difference between two temporal categories lies in the fact that the category of tense in English is a four-member one, according to the number of the axes of orientation, and it represents a vertical cut among verbal forms, while the category of temporal relativity is a two-member one and is situated on the horizontal line of each time sphere; it lacks the stage of succession of the action and has only a contrast of simultaneity–anteriority, directly related not to the moment of speech but to any one axis of orientation.

Besides, one category is contained within the other — perfectness depends on tense and is included into it, and not vice versa.

The morphological formants of these two categories are also different, perfectness being marked in a number of perfect forms. The category of tense reflects coincidence with the axes of orientation, synchronization of the action in regard to the axes, while the basis of perfectness is non-coincidence of the action with any axes, it is anteriority to them.

We think it interesting to present the axes of orientation according to Bull's scheme:[1]

where *RP* is retrospective point, *RAP* retrospective anticipated point, *PP*, prime point, *AP* anticipated point.

These axes, which we mark by the usual terms of 'present *(PP)*, 'past' *(RP)*, 'future' in two aspects, *F* I *(AP)* and *F* II *(RAP)*, mean really existing grammatical relations, the referent of which, indirectly, are temporal relations of reality.

What sets or paradigms are direct exponents ot the grammatical category of tense? The formal indicator of relation of tense (i.e. to some axis of orientation) is present in all the four paradigms of each set, corresponding to a certain axis of orientation. However, in the sets (b), (c), (d) (continuous, perfect and perfect-continuous) the relation to the axis of orientation is but ground for comparison of relations of these sets within one common temporal plane, it is a temporal basis on which are revealed grammatical oppositions in other categories within the given axis of orientation. The temporal marker in these sets coexists with markers of other categories (aspect and perfectness), while in the indefinite set on all the axes of orientation the basic feature of all the forms is the relation to the category of tense, which is not complicated by any other categorial features or relations.

The indefinite set can be an unmarked member of oppositions in other categories precisely because it retains, as the only categorial marker, the formant of tense, which serves as ground for

1 Bull (1960) 26.

comparison within other members of the set of the given temporal plane, the members being linked with the indefinite set by different categorial contrastings.

The indefinite set is neutral to the category of aspect (duration in English), so its treatment as a common aspect, present in some grammars[2], can hardly be accepted, since there can be no such aspect; 'common' is not an aspectual feature but simply an indication that there is no aspect at all. This term tells of the unmarkedness of the weak member of a privative opposition in aspect (duration). The same term is valid for case ('common' case as opposed to possessive in nouns) because the relations between these two cases are not privative, as with the verb sets in question.

What invariant meaning is contained within grammemes of the indefinite set?

The invariant content of the grammatical form of the present indefinite is the reference of the action (event) to the temporal sphere, somehow or other including the present moment; it is an action valid for that moment, simultaneous with it, the specific nature of the moment of speech itself making the present spread in the direction both of the past and into the future. That is why *Pr.I* may denote a habitual action, or some established rule, valid for all temporal stages or, in a narrower sense, some action simply fixed at the moment of reporting it, taking place at the present time. The use of *Pr.I* in so-called stage directions is justified by its invariant meaning, and it is wrongly considered as belonging to stylistical cases of using *Pr.I.* In the sentences "*Do you take sugar, Mr. Cokane?*" or "*You are very welcome*",[3] we have the reference of the action to the moment of speech of the play's character; the first sentence denotes a habitual action not limited narrowly by the moment of speech, while the second action is limited by the moment of the utterance. The range of the present is practically infinite, but a necessary condition of such an action is its existence

[2] Barkhudarov and Steling (1960) 160.

[3] Literary examples are taken mainly from the plays by Bernard Shaw. In other cases the author is indicated.

in one plane with the moment of speech, the inclusion of the moment of speech into the sphere of the action, not anteriority or succession to it, and with the moment of speech itself excluded. *Pr.I* is widely used in narrations taking place within the sphere of the moment of speech (especially in plays).

Stage directions are worded in *Pr.I* because they show synchronization with the speech of the characters in the play. Some more examples are *He is overcome afresh and the rest collapse again. A sudden darkness blots out the walls of the room as they all spring to their feet in consternation.*

From the temporal qualities of the moment of speech follows the ability of *Pr.I* to express the characteristics of an object (event), its permanent qualities, which seem to be valid at all times, to express universal truth, as, e.g. *Meat is very nourishing. Water freezes at zero. I am a student. Old people become peevish*, etc.

This also implies the possibility of denoting a repeated, usual, habitual action: *Do you play bridge? They dine with us. What do you teach? He goes to a night school.* The action may be momentary: *Who goes there?*

All the individual meanings of *Pr.I*, syntagmatically conditioned as its variants, in its use to express momentary or iterative actions, universal truth, or the subject's ability to perform some action, may be reduced to the general invariant of the action as a fact, related to the moment of speech, simultaneous with it, and neutral to duration. Examples where the action coincides with the moment of speech are *I beg my readers not to mistake my journalistic utterances. I regret the mistake, sir*, etc.

But, as is known, there are instances of using *Pr.I* where it acquires the meaning of actions past of future. Can such cases as 'the historic present' (as O. Jespersen calls it more adequately the 'unhistoric present') be explained by the same invariant meaning of reference to the moment of speech? Certainly yes, but the reference of an action to the moment of speech may be realized only through the process of TRANSPOSITION, i.e. the transference of the action into the range of another axis of orientation. The transposition of an action into the plane of the moment of speech of the speaker, who, in certain contextual situations, transposes a past action into the plane of his speech, easily takes place. At the basis of 'historic present' lies an event of stylistic order, which has become systemic and fixed as a language norm, and is not an accidental phenomenon arising in some contexts on an individual stylistic basis.

It is noteworthy that transposition takes place not from the present into the past but vice versa, since *Pr.I* is the initial member of temporal oppositions, containing a more general and wide meaning.

It is not accidental that such a more general member is precisely the axis of orientation of the present, since the present is the moment of speech, a fact that logically justifies the attraction of different actions into this sphere. In these usages is shown a wide range of the sphere of the present, its ability to spread in retrospective and prospective planes.

The transference of a FUTURE action into the moment of speech occurs through neutralization of the PLANE OF EXPRESSION of the future action, which is possible if a lexical marker of the future action is present in the sentence, while the grammatical marker, then, becomes superfluous and the meaning of the future is transferred onto the adverbials. As is the case with neutralization generally, in the position of neutralization appears the member with the most general meaning and wide range, in such examples as *I leave the place tomorrow*, the universal quality of the present allows its use for the expression of a future action. This is facilitated also by the fact of a closer tie of the future action with the present one, than, say, the tie between the past and the present, mainly due to the modal nature of the future action, which is not tangible and cannot be observed and, thus, is projected from the present. This allows for neutralization of the future both when using the present to express a planned action (in most cases a planned movement in space, when the verb is usually accompanied by some adverbial modifier, implying futurity), and when the present is used in subordinate clauses after a number of conjunctions, also designating the future in their content, or else after modal conjunctions, *tell me when he comes: If I find that you have stepped an inch outside the letter of the law ... I will prosecute you myself.*

Neutralization of the plane of expression (form) of the future is also seen in the sentence "*Well, what do you say now, Mr. Sartorius?*"

These cases of neutralizing the future also have their roots in the historical process of the development of means denoting future actions, since such actions are known to have sprung from the sphere of the present. The final fixing of special formants, modal in their nature, actually refers to the start of the modern period of English language development. The above fact also enables us to consider neglecting the plane of expression of the future as an instance of the neutralization of this plane in conditions which make the planned nature of the action clear from the distributional data. In this connection it would be of interest to compare the instance of transposing the past into the plane of the present and the use of present instead of the future. These are different phenomena. The second case contains no transposition, since the present does not transpose the future action into its temporal plane and itself is not transposed anywhere; this is confirmed by the presence of adverbials, indicators of time, while in cases of historic present there are no indicators, and the action, by the will of the speaker, is actually transferred completely from the sphere of the past into the present, the speaker bringing it closer to himself. Hence, we have here neutralization of only the plane of content of the past, of a stronger member of the opposition 'present–past', which occurs through transposition, while in the second case the FORM of the future is neutralized, also a stronger member of this second temporal opposition. This is the instance of the present spreading into the sphere of the future, which is mentioned by A. Tellier.[4] Speaking of the general invariant meaning of *Pr.I*, we can hardly agree with the point of view that the present is devoid of any semantic content;[5] it is

[4] Tellier (1962) 308.
[5] Togeby (1953) 29.

true, however, that it is a standard from which we reckon,[6] a weaker member of the opposition.

When analysing the grammatical meaning of a grammeme we should differentiate between the heterogeneity covered by the invariant, and some additional meanings which the form can acquire due to some special processes such as neutralization, or in specific syntagmatic or stylistic conditions. However, the stylistic variants of the meaning may also be explained on the basis of the invariant, which is always present. Different stylistic uses are conditioned by the the invariant, since otherwise they could not be motivated, and there can be no unmotivated meanings and uses. The stylistic purpose of a unit usually is to stress a certain aspect of its general meaning, which, however, is preserved.

In this connection it is possible to establish those variants of the meaning of *Pr.I* which are included into the invariant and those which the form may acquire under special conditions. Thus, for example, the ability of *Pr.I* to express an action occurring at the moment of speech, a habitual action,[7] or universal truth, all this refers to the invariant and makes up the semes of the form of *Pr.I*, since the action is throughout valid at the moment of speech. On the other hand, the meaning of a past or future action cannot be considered as one of the semes (components of meaning) of *Pr.I*, since these actions are not happening at the moment of speech, and consequently are not covered by the invariant. They may be motivated by the invariant from the point of view that the speaker can transfer any action into the sphere of his speech. Hence, the present cannot have a seme of the past, just as the past cannot have the seme of the present, etc. They are but specific syntagmatically-conditioned instances of use, and not components of the meaning of the grammemes.

Revealing the processes, reflected in the content of a grammeme

[6] Calahan (1938) 146. Cf. The division of tense into general *(allgemeine)* and past: Glinz (1952) 109.

[7] Sten (1952) 12 admits that when the present expresses a habitual action, the moment of speaking is not expressed, but the present moment is implied during the whole of the iterative action.

is of great practical importance, since the explanation of the cause and motivation of different meanings of one form greatly facilitates its study.

It should be kept in mind that when we speak of neutralization of the plane of expression or content, when in the position of neutralization a more general member appears, viz., the present, we mean the possibility of neutralizing the temporal plane not only in the forms of the indefinite set (which is a pure exponent of the category of tense not complicated with any aspectual *DF*), but in other forms of the same temporal plane as well. For example, *Pr.C* as well as *Pr.P* may be used in the meaning of the future. These are syntagmatically conditioned meanings of the forms of *Pr.C* and *Pr.P* of the type *I am leaving to-morrow*, or *After I have received the letter, I shall come to you*, or *It was twelve and the train was leaving at six.*[8]

The invariant content of the PAST INDEFINITE *(Pt.I)* is its isolation from the present, the action belonging to the sphere of the past. The past means anteriority to the present when this anteriority is of a purely temporal nature (not relative), creating a special temporal axis of the action, preceding the moment of speech and passing onto another, special axis of orientation, and hence, not linked with the present. The isolation from the present testifies to the action being on another temporal axis and having no ties with the axis of the present, thus defining the syntactic interdependence of the verbal sets of the past sphere, which usually do not mix with sets of the sphere of the present and future on the syntagmatic axis, in other words, in speech.

The axis of the future, although it is also an independent temporal axis, shows a closer contact with the axis of the moment of speech. The closed character of the axis of the past on the one hand, and of the present and future on the other, facilitated the rise in the English language of a special axis of future II (relative) and, hence, of the form it represents. Thus, in such languages as French and English there appear specific syntactically dependent types of

[8] The example is taken from Diver (1963) 168.

future, denoting a planned action, proceeding from the narration in the past.

Evidently, a clearly expressed isolation of the past from the present in English is due to the fact that the expression of anteriority, parallel on all axes of orientation, has been firmly fixed in the language.

The perfect of the present tense has undergone various changes in different languages during their development. This shifting of its meaning may be described as the development from the meaning of the result of a past action, stemming from its inner form, further to anteriority to the present, and then further to the past proper. Such development is quite logical because anteriority to the present is easily mixed with the past proper, which also reflects anteriority to the present but on a different basis. This happened in the German language where *Pr.P* can have both the meaning of the past and the perfect. In languages where the meaning of anterior-present, parallel to anterior-past and future has appeared, the past became isolated, and then conditions appeared for the development of the relative tenses of the sphere of the future (in French and English). Besides, the relative temporal meaning of perfect has created a symmetrical frame in the whole verb system; it has helped to fix the prevalence of tense over aspect and created the soil for developing such specific aspect forms as the durative ones. This frame of the proper temporal and relative temporal divisions supports the system and makes it possible to fill all its empty cells; this has happened in the English verb.[9]

Thus, among indefinite sets, placed on different temporal axes of orientation, there exists a correlation of a purely temporal plane and, hence, they form the basis of the grammatical category of tense. The general, invariant content of the whole indefinite set corresponds to the temporal content of each one axis of orientation. Separate cases of use, including those conditioned by special conditions of distribution, somehow or other confirm their reference to some temporal plane.

[9] The question is discussed in Khlebnikova (1966a) 62-73.

It is because the indefinite set serves as a formal exponent of the grammatical category of tense, that its grammemes are devoid of aspect differentiation; in fact, an action defined in indefinite formtypes may express different shades of duration and completeness, all in accord with the specific nature of the temporal plane it refers to. Such a purely temporal nature of verbal forms, not complicated by any special denotation of the character of the action, distinguished the verb in the Old Germanic languages. The shades of aspect which the temporal form may acquire in Modern English are not conditioned by the categorial meaning of this form, but are linked with the nature of the reflection in the lexical meaning of separate lexical units, of the terminative or non-terminative meaning of a particular verb, or may depend upon the immediate environment of the form.

The isolation of the axis of the past from the moment of speech and non-contact with other actions, makes for a specific use of *Pt.I* and excludes from its sphere everything that is most typical of the panchronic present — the denotation of permanent qualities, characteristics of events, and universal truths which are valid generally. Hence the main sphere of the use of *Pt.I* is the narration, the representation of a chain of isolated actions or separate isolated actions which should be necessarily prior to the moment of speech. The nature of the process of the action finds no reflection at all in the categorial meaning of the *Pt.I* formtype, and a verb of any individual aspectual content may be used in this form.

The specific nature of the two future indefinites also fully corresponds to the nature of their temporal planes. They denote a planned action, anticipated with a sufficient degree of confidence, the difference between the two planes of future being not so much categorial as syntactic, conditioned by the presence in the environment of forms of the sphere of present or past: *F* I and *F* II are distributionally defined varieties of the future indefinite; their forms reflect their grammatical destination, a factor which cannot be disregarded, because every morphological form has its own categorial meaning and place in the microsystem. It is interesting to observe here that the categorial content of both varieties of the future is analogous: any action expressed through a *F* I indefinite form may be transferred into the sphere of the past and substituted by a form of *F* II indefinite without any change in the categorial content of the verb. The similarity of the invariant meaning of systemic grammemes is unusual in morphology; synonymy does not find a place in the morphological network of a language. How then do we have the right to proclaim systemic properties of both *F* I and *F* II if they have the same general meaning of futurity? This problem, alongside with many others, finds solution in the componential analysis of grammemes into *DF*. In this respect the peculiarity of both *F* I and F II grammemes manifests itself in the combination of *DF* within their invariant meaning: in *F* I the *DF* of future is combined with that of present, and in *F* II with that of past, the latter two semes being subordinated to the main one (future). That is why between themselves the two futures stand in opposition, having for the basis of comparison the *DF* of future, and the basis for distinction the morphosemes of present and past. The community and distinction of both futures can be easily illustrated in any context. Thus, e.g., if the following phrases from B. Shaw should be preceded by a narration in the past, then, except for a process of transposition into another temporal sphere, nothing will change in the content of the actions themselves: "*Then I will drive this pilum through you*" = *he said he would drive that pilum through him.* "*I shall not bother about them*" = *he said he would not bother about them.* "*You will keep faith?*" *She asked her if she would keep faith.*

The panchronism of the action, natural for the sphere of the present, is unnatural for the future (as well as for the past). This fact illustrates the specific character of each temporal axis of orientation. The varieties of future are not only in opposition to each other, but also to the axes of present and past respectively, since they are contrasted on the line of different temporal planes. The opposition in meaning is supported by the opposition in form, in which these temporal planes are reflected. It is the complete equivalence of the main categorial *DF* of *F* I and *F* II that leads to the wide use of *F* II in Modern English, not only in indirect speech and complex sentences, but also in any context of the sphere of the past denoting future actions set on the background of the temporal plane of the past. The very fact of syntactic dependence of all tenses of the sphere of the past makes *F* II a component of a complex sentence to a greater degree than *F* I, similar to the Past Perfect being linked to a greater extent with the subordinate clause than is the Present Perfect. All these and other syntactically defined dependences are based on the nature of the temporal plane with which any one form of tense-aspect is linked.

As to the problem of the modality of the future, it also, in the final count, is connected with the nature of the temporal reference of the action, since it is natural that any action which does not and did not occur, but is simply anticipated, in its very anticipatory nature is modal in content, and, depending on the degree of supposition implied by the speaker, it may be denoted either in forms of future indicative or in forms of subjunctive mood *(SM)*.[10] This is where the problem arises of the boundary of future and *SM*, especially in English, where the presence of *F* II brought the future even closer to *SM*, which was facilitated by the homonymy of their formants. It is the modal nature ot the axis ot orientation of the future that accounts for the possibility of choosing formants of this tense from the inventory of modal verbs expressing an action desirable or necessary. The community ot the formants of future and *SM* in Indo-European and especially in Germanic languages is due to their modal proximity. The actual data on Germanic languages given in G. Sčur's article,[11] show that in all Germanic languages the auxiliary verb of future tense and *SM* is the same, the only difference being that in languages where inflectional means of expressing *SM* still exist, this verb shows the mood in formbuilding elements (cf. German *werde schreiben–würde schreiben*). These facts are also proof of the validity of one of the principles of structural organization of verb systems

[10] Without going deep into the problem we shall apply to both the subjunctive and conditional forms the term 'Subjunctive mood' *(SM)*.
[11] Sčur (1964) 49-57.

— structural uniformity of categorial forms in different microsystems; they also testify to the presence in the microsystem of tense-aspect of all those formants which are found in other microsystems.

Into what kind of oppositions can grammemes which express the category of tense be joined? Previously, the question of the nature of oppositions in the verb was considered in chapter six, where a thesis was put forward of the equipollent nature of temporal oppositions. As was stated above, the English *Pr.I* (like the present tense in other languages) marks a definite *DF* — the seme of inclusiveness of the moment of speech. *Pt.I*, on the contrary, signals exclusiveness of the moment of speech, anteriority to it, while the future is carrier of the temporal notion of succession to the moment of speech, of an anticipated action, all these qualities of temporal forms being marked by definite formants. Thus, everybody knows that *I love*, *he loves* expresses present, *I loved*– past, and *I shall love* — future. Hence, it is hardly possible to speak of privative oppositions to be found in the contrasts of temporal forms. Maybe there are no oppositions here at all, and each temporal form is only entering as a component into some other oppositions. Temporal sets certainly form oppositions, and binary at that. In languages with a triple scheme of temporal divisions, oppositions ? consist of two pairs, one with the action directed from present into past, the other — from present into future.

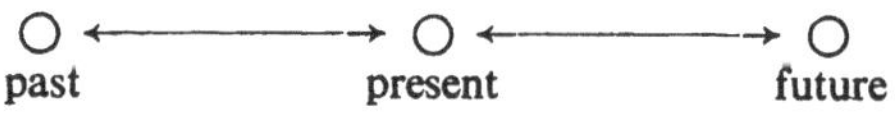

These two directions from the moment of speech are not equivalent. If the past and present form really polar notions, not adjoining each other without interference of special processes (e.g. transposition), the present and future are closer to each other, the future stemming from the present. The contrast on the plane of tense, therefore, may be shown in the following way as well:

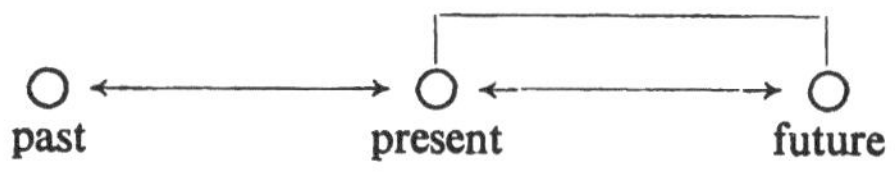

where the opposition present-future is inside one member of the general opposition past-present. The nature of relations among these three members makes it possible to neutralize the past only through transposition in the plane of the present, on the one hand, and creates the chance, on the other hand, to neutralize the PLANE OF EXPRESSION of the future, precisely because future is closely linked with present.

Temporal oppositions are equipollent since their members differ from one another in temporal *DF*, and they enter contrasted distributional frames, which reflect that distinction; thus, e.g. the sentence, *I shall come to you yesterday* is inadmissible. The basis for comparison in these oppositions is the relation to the present moment (the act of speech). In the above scheme both oppositions are built on their relation to the moment of speech, while in the opposition past–future II, introducing an additional temporal axis of orientation into the scheme of temporal divisions of the English verb, the basis for comparison is the moment of speech in the past, the starting point of the whole opposition, the first member denoting coincidence with this moment and the second posteriority to it. But as distinct from privative opposition, where the extensive member is devoid of the sign asserted in the second, intensive member, in the equipollent opposition the initial member marks the *DF* which is the basis for comparison, and in the second, stronger member the contrasted *DF* is marked. The member, which marks the initial *DF* is more general, universal, initial. The *Pr.I* is such a member; it is the initial member of two oppositions in tense, and, as will be seen further on, in other categories as well. In a privative opposition the basis for comparison remains one of the *DF* within both members of the opposition, while in the stronger member of the equipollent opposition this *DF* (e.g. the present) is not marked at all; it is denied by the very fact of markedness of another, new *DF*, which in the long run exists only by way of contrast to the initial *DF* which serves as a basis for comparison in temporal oppositions. This is the gist of the equipollence of temporal oppositions, which is verified in the relation to neutralization, thus proving the essence of the hierarchy of opposed members.

The context also illustrates the presence of contrasting, but mutually linked relations of two members: thus, e.g. the sentences *I know this man–I knew this man before* show different relations to the moment of speech, just as *I speak English–I shall speak English soon*, as well as the pair *I know he will learn to speak soon–I knew he would learn to speak soon*, which shows the contrast in temporal marking of actions, considered from different temporal points.

Equipollence of oppositions does not mean a complete equivalence of its members, which might make doubtful the very fact of the existence of the opposition; it implies a relative non-equivalence of the members in their mutual markedness.

Since temporal oppositions are not privative by their very nature, their behaviour in the process of neutralization reflects their equipollence. The point of view that the process of neutralization in its very content is characteristic only of privative oppositions, stands the test of temporal oppositions, for one can hardly consider as true neutralization the process of transposition of the past into the sphere of the present in cases of historic present. This process leads rather to polyfunctionalism of the grammeme *Pr.I*, and not to syntagmatically conditioned suspension of the relevance of the content of one of the counter-members. The case of neutralization of the opposition present–future is the neutralization of expression, and not content, i.e. it is not a complete neutralization (as in privative oppositions).

Thus, oppositions in the grammatical category of tense are binary and have *Pr.I* as the starting point in which these oppositions intersect. The opposition past — *F* II is, in fact, a syntagmatically conditioned transposition of the opposition present — *F* I into the plane of the past axis of orientation. The two grammemes denoting future, *F* I and F II, are in opposition as well, for this opposition also reveals the contrasting qualities of these two members, of future actions, proceeding from different temporal planes.[12]

[12] A similar understanding of the correlation between simple future and

The general content of indefinite forms is the localization of the action in tense (grammatical), without introducing into the action any other categorial meanings, such as aspectual, modal, or voice *DF*. The categorial content of the members of temporal oppositions is marked in special morphological formants in all the members of temporal oppositions. The zero formant of *Pr.I* is a formal indicator of precisely this formtype in contrast to other members of temporal oppositions; in the third person singular, the zero ending gives place to the exponent of present *-s (speaks)*, which is also the formal sign of this tense.

A similar temporal contrast is preserved in other grammemes on different axes of orientation, but they are complicated by semes of other grammatical categories and, therefore, are not purely temporal forms. Thus, e.g., *Pr.C* is contrasted to *Pt.C* on the basis of tense, but the ground for comparison in them is the sign of duration, common for both forms.

Since the invariant of each grammeme is made up of a totality of semes, relevant only for it as a member of some categorial opposition, it would be appropriate to define for each set the composition of semes which give a start to oppositions. And, since the nature of oppositions is not uniform (homogeneous) and the members of oppositions are not equivalent, but form a hierarchy, the degree of relevance of different semes is different in each grammeme. Therefore, it is necessary to distinguish weak semes, which serve as a latent unit of contrast and strong ones marking the *DF* relevant for the opposition.

As temporal oppositions are equipollent, temporal semes in all the formtypes are strong, while the semes which may be singled out within them only as in an unmarked member of the opposition are weak. Thus, e.g. in *Pr.I I write* three semes may be singled out (in all the grammemes of the microsystem of tense-aspect three semes are present, shown in the scheme of intercrossing oppositions on page 157); the strong seme is that of present, serving as *DF* of the category of tense; the weak semes are those

future-in-the past with reference to French can be found in Camara (1964) 547-551.

of the unmarked member of the aspect opposition and of simultaneity, which merges into one common *DF* with the seme denoting the axis of orientation; it is evident from the formant of the present tense that simultaneity to the given axis is expressed.

The same is true about *Pt.I*, as well as F *I* and *F* II, with a common invariant of the future. In the formants of two future tenses are combined the *DF* of present (*F* I) or past (*F* II) and future *(will do–would do)*.

The strong seme is always reflected in the formant. So, since equipollent oppositions are equivalent, all their members have a formal marking of the temporal *DF*.

12.

OPPOSITIONS IN THE CATEGORY OF TEMPORAL RELATIVITY (PERFECTNESS)

The next set of correlations, which may be traced in the scheme of tense-aspect, is the contrast of the sets (a) and (b) (non-perfect) on the one hand, to the sets (c) and (d) (perfect) on the other. On what invariant meaning are oppositions in temporal relativity (perfectness) built?

The general meaning of the grammemes of the category of perfectness is based on the nature of that axis of orientation the relation to which is reflected by these grammemes. If the exponents of the category of tense include one relevant categorial *DF* — tense (the specific nature of some axis of orientation in regard to the moment of speech), then the exponents of the category of perfectness, due to their reference to a particular axis of orientation, include two relevant *DF* — the *DF* of the temporal axis of orientation itself on which the opposition in temporal relativity is placed, and the *DF* of the NATURE OF RELATION to the given axis, simultaneity or anteriority to it. This makes for the difference between these two different types of oppositions, although they are temporal in their basis. The direct exponents of the grammatical category of perfectness are the grammemes of the indefinite and perfect sets which are not complicated with *DF* of duration. However, the indefinite set, possessing a more generalized meaning and marking only the temporal axis, may enter other oppositions in different grammatical *DF*, being their unmarked member in the plane of expression and in the plane of content. Can we consider the indefinite set to be an unmarked privative member of a relative temporal opposition, and the whole opposition of non-perfect–perfect forms as a privative one? Evidently not. The fact is that in a privative opposition the weak member, unmarked by the *DF* of the strong member, according to the logical basis of such an opposition, should contain the basis for comparison, which is necessarily present in both members of the opposition. Simultaneity, however, is not present in the second member, the perfect set, which retains as a basis for comparison only the *DF* marking its belonging to a certain axis of orientation, this *DF* being common for any members of the opposition on one temporal axis (e.g., the formant of the present is active in any grammeme of the temporal plane of the present — *have done, is doing*). The other marked *DF*, found only in the strong member, is that of perfectness (priority). That is why this opposition is termed here as an intermediate one between the privative and equipollent (cf. chapter 6) in which simultaneity, though formally unmarked, is a relevant *DF* of the plane of content for the given opposition. Thus, the indefinite set, as the one with the most general meaning, becomes the initial member of another opposition. The fact of the existence of this opposition is confirmed by a clearly expressed content of anteriority in the perfect set, in contrast to the meaning of synchronization of the action with the given temporal plane in the indefinite set. The definition of the content of all perfect grammemes in all microsystems of the verb as the content of anteriority is based on the principle of balance inherent to the system, another principle of structural organization of the verbal system which forbids the existence of heterogeneous content and expression in homogeneous, proportional sets of verbal forms. It is necessary to show now to what extent the above principle is confirmed by the language material of the tense-aspect *MS*.

In the perfect set a strong seme of anteriority is easily detected (or perfectness — for the English language perfectness is a symbol of anteriority) and it is formally marked, then a weak seme of the counter member of the opposition in aspect (duration), and then the seme of the temporal axis, which serves as a basis for comparison of this opposition placed within a definite temporal axis. Since the last seme has a formal expression, it cannot be called weak, though it is less strong than the one serving as a basis for distinction of the members of the opposition.

Semes *(DF)* form a hierarchy of relations, since otherwise they would not be the basis of oppositions.

It is common knowledge that a great number of conflicting views exist in connection with the grammatical content of the present perfect *(Pr.P)*. Proponents of the aspect content of the form ascribe the following meanings to it: result, completeness, retrospectiveness, continuity of the link between past and present, etc., it being noteworthy that such meanings as retrospectiveness and continuity actually are not aspectual categories, but convey the content of anterior present, i.e. of temporal relation in the plane of anteriority. The difficulty of the situation, arising when attempts are made to impose the meaning of aspect onto the perfect, is well revealed in the concept stated by A. Marckwardt, when, admitting the ability of the perfect to express completeness of the action, Marckwardt cannot accept the aspectual nature of the perfect, since, as he rightly points out, the indefinite set can also express completeness, that is, both members of one opposition denote the same thing.[1]

It is worth noting that H. Sweet, a progressive grammarian of the older generation, detected the meaning of anteriority in the perfect.[2] It is quite possible that those linguists who share this point of view are, to a certain extent, influenced by him. The content of *Pr.P* as well as other perfect sets cannot but depend on that temporal plane on the basis of which is built the opposition non-perfect — perfect. The specific nature of *Pr.P* is also due to its containing a certain contradiction, which consists in the following: the past is also anterior to the present, and this means that there are two anterior present tenses — one represented by a simple past, the other by *Pr.P*. However, such a situation may arise, although not necessarily in all languages, due to the

1 Marckwardt (1942) 134.
2 Sweet (1892) 275-281.

systemic nature of verbal forms, which manifests itself in proportional and binary oppositions. The presence of a relative form in the plane of the past implies its presence in the parallel temporal plane. If we admit the existence of oppositions in tense among the axes of orientation, then we should also admit that all the sets of each temporal plane are in opposition to the parallel sets of the next temporal plane, and that the general categorial content of perfect, indefinite, and other sets is equal on different temporal axes. It should not be forgotten, however, that anteriority to the moment of speech as a relative category in the conditions of the existence of a set, expressing anteriority to the moment of speech in a purely temporal plane, in forms of *Pt.I*, cannot but find reflection in the categorial meaning of *Pr.P.* If *Pt.I* expresses an action isolated from the present, since it is placed on a different temporal axis, then *Pr.P*, as a form belonging to the axis of the present, expresses a contact anteriority, somehow or other linked with the present. On the other hand, it is the categorial meaning of anteriority that lends to the relative temporal form the meaning of completeness, since, if the action precedes the moment of speech, it should be finished, stopped before this moment. Such a contradictory character of the meaning of anteriority, not stemming from the past, but linked with the present, creates an increased sensitivity of *Pr.P* to the lexical meaning of the verb, and leads to some cases of confusion and close contact of *Pt.I* and *Pr.P.* These points of contact have also been studied by many linguists.[3] Compare, for example: *Did you tell me everything?* and *Have you told me everything? Did you see all the patients?* and *Have you seen all the patients?* The means of studying contiguity and distinctions among these formtypes lies in the close analysis of the lexical filling of the forms, as well as their distributive data. *Pt.I.* and *Pr.P.* can express homogeneous actions (as in the above examples). The choice of the form in such

[3] Bryan (1936) 363-382. The comparison of preterit and perfect can be also found in the book Dietrich (1955) 137 ff., 183. The author draws attention to the importance of the point of view from which the speaker regards a past action starting from the present or from the past.

cases depends on the point of view from which the actions are considered — as isolated from the present or as linked with it. As H. Reichenbach rightly points out, in *Pt.I* the point of the event and the point of reference coincide, both preceding the moment of speech, while in *Pr.P* the event precedes the moment of speech but refers to the point simultaneous with the moment of speech, i.e. the point of reference and the moment of speech coincide. Thus, the past events are considered in *Pr.P* as proceeding from the temporal point, coinciding with the moment of speech. This fact, according to H. Reichenbach, explains why *Pr.P* is not used in narration, but creates the impression of information conveyed to the reader.[4]

It should also be stressed that the temporal reference of *Pr.P* contributes to the fact that in direct speech, which by itself implies the coincidence of the point of reference with the moment of speech, all the more so when surrounded by the forms of the sphere of the present, the perfect may mean a past action, except in combinations with clear lexical markers of isolation of an action from the present when the *Pt.I* must be used. The very fact of including the action into the sphere of direct speech testifies to its reference to the present, as in the example "*Have you brooded?*" — "*Yes, sir*" — "*With what success?*" — "*I have a plan, sir*".[5] If we are to admit the exclusively aspectual meaning of the perfect, then it is impossible to explain its contiguity to *Pt.I.* This contiguity may be explained only on the basis of the relative-temporal meaning of the perfect, since it is exactly relative anteriority to the present that can coincide with temporal anteriority, expressed by the axis of the past. The recognition of the aspectual function of the perfect set is hampered also by the fact that aspect, if it really is a grammatical category, should be completely abstracted from the lexical characteristics of the verb and should be evenly spread on all axes of orientation. The absence of a categorial aspectual *DF* even in just one case suspends the possibility of examining the given form as an exponent of aspect, all the more

[4] Reichenbach (1947) 288-9.
[5] Voroncova (1960) 99.

so if there is no correlative pair to this aspect. It is quite evident, however, that no aspectual meaning can be found in the perfect sets of the gerund, participle, subjunctive mood, and even past and future perfect.

Different semantic shades of the meaning of actions expressed by the form of *Pr.P* are due, first of all, to the lexical nature of the verb, its terminative or non-terminative meaning. The reaction of perfect forms to the terminativeness of the verb is increased in the temporal plane of the present owing to the peculiarity of such a conflicting combination in the *Pr.P* as that of the plane of the present and the relative temporal plane of anteriority.

Thus, verbs of a neutral lexical nature used in the perfect acquire the meaning of a completed action (not to speak about terminative verbs in which perfectivity is added to anteriority, the form acquiring the meaning of a completed action), e.g., "*Think a moment what you have said!*"

Non-terminative verbs also yield to the influence of the form, but, as if resisting this influence, they may acquire the meaning of an action, although an anterior present one (and, hence, completed), but nevertheless continuing at the present moment or completed on the threshold of the present. In other words, the perfect of such verbs acquires a meaning equivalent to the meaning of the perfect continuous. For example: *I have had a good deal of experience guiding young people. I have kept you all this time for nothing*. Such examples represent pure cases of neutralization of the opposition *Pr.P–Pr.PC* under the influence of the lexical meaning of the verb *(have)* or some unit of the context *(all this time)*. In the position of neutralization there appears the privative member of the opposition in which duration is unmarked.

In the past perfect, due to its seclusion within the frame of the past, a lesser degree of dependence of its content on the lexical nature of the verb can be observed. Anteriority to the plane of the past imparts to the action the content of the pluperfect, making it possible for this formtype to be used in narration. As an example we may cite almost a whole page of *Pt.P* forms from J. Galsworthy's novel *In Chancery*: *For long years, Soames' acqaintanceship with Soho had been confined to its Western Bastion ... Many bargains had be picked up there ...*

The meaning of anteriority also explains the use of *Pt.P* to denote a momentary action, completed by a certain moment, fixed either by an adverbial modifier of time or by any other action, or by the general context; (1) *Inside my room the tree I had set up weeks ago for the New Year had shed most of its yellow blossoms. They had fallen between the keys of my type-writer.* (2) *I suddenly remembered that I had seen her.*

As to the future I perfect (*FIP*), and the more so *FIIP*, the nature of their temporal reference influences their use; however, anteriority to the future is practically a rare case. These forms simply illustrate the redundancy of the system and only potentially exist in the language as members of binary oppositions which fill the cells of the *MS*. In any case the perfect set in all temporal planes has a general categorial meaning of anteriority to the given temporal plane (axis of orientation) and is a purely relative tense.

The treatment of *Pr.P* and all perfect forms generally affects the construction of the whole system of tense-aspect by analysts, and it is the recognition of aspectuality as a categorial *DF* of the *Pr.P* that breaks the system, deprives it of its harmony, and makes it disorderly, the latter quality being not characteristic for morphological sets of any language, all the more so for the system of English tenses, which is deservedly thought to be ideal in its symmetry. If the continuous set is in most cases interpreted as aspectual in the works of later years, then the perfect-continuous set reflects the approach to the perfect, thus once more confirming that the perfect set serves as a temporal frame for the perfect-continuous one. An attempt to join two different aspects into one form (and this is the natural result of interpreting perfect as a form of aspect) is clearly inconsistent, since there is still no language in which two different aspects or tenses might be joined into one form. That is why, for example, I. Ivanova has to admit that "the perfect-continuous set does not possess an aspectual meaning which would correspond to the aspectual meaning of the perfect", and "the presence of similar traits of two mutually opposite sets forces us to admit that the perfect-continuous set is an independent category to be analysed independently and not under the heading of those sets with which it has common points. If it is so, then the given form evidently has its own inherent grammatical meaning".[6]

This meaning is found by the author through testing the reaction of verbs of different lexical character, i.e. by means of a lexical category, and not a grammatical one.

To confirm the relative temporal meaning of the perfect set, here are some examples: "... *I find her under the impression that something has passed between you which it is my duty as a father ... to inquire into at once. My daughter ... has taken you quite seriously ...*", "*Are you sure Dick has not been touched?*" "*I have merely taken a furnished villa at Surbiton for the summer.*" "*I presume you have been round this little place. Not much to see here ...*", etc. *Pr.P* shows an action which is past in its relation to the moment of speech and not necessarily linked with it by its result. The action is connected with the moment of speech only because through its formants it refers to the sphere of the present, as well as because there exists the *Pt.I* which conveys an action cut off from the present as in "*I had no idea that the architect of Cologne cathedral lived so recently.*"

If the context of speech is transposed into the plane of the past, then *Pr.P* may be substituted by *Pt.I*, and no 'special aspectual content' will be changed in this case. It is the point of view from which the action is considered that is of importance. That is why *Pr.P* may often be substituted by *Pt.I* without any significant change in the sense of the sentence. Thus, in the sentences "*I have been a young man myself ...*", "*He has been too well brought up by a pious mother to have any sense or manhood left in him*" *Pt.I* may be used instead of *Pr.P* in the corresponding environment, and the meaning of the action will remain unchanged. But the very fact of speech taking place in the present makes it more usual to use the tenses of the sphere of the present. The connection

[6] Ivanova (1961) 147-149.

with the present makes an action expressed by *Pr.P* valid for the present, thus creating the content of a resultative action often enhanced by the lexical meaning of the verb: "*But I fear I have disturbed you in the composition of your letter.*" "*We have secured the room with the best view in the hotel, Harry*". "*And now, Dr. Trench, since you have acted handsomely, you shall have no cause to complain of me* .."

The specific nature of the temporal plane, preceded by an action expressed in *Pr.P*, creates all its meanings, which are usually considered its quality, such as the relevance in the present of events taking place in the past, resultativity, and the connection of a past action or state with the current event, etc.[7]

Pr.P may be used as a member of the neutralizable opposition in duration, in instances when the action spreads within the limits of the present tense. Such a neutralization is often lexically conditioned and, hence, is constant, as is the case with the verb *to be*, which cannot have any forms of perfect continuous: *I have been uneasy all the afternoon about what they must think of us. I have been hungry all day.*[8] Neutralization can also be found in the following: "*Papa has not been disagreeable, I hope.*" The example below shows neutralization of the plane of expression of a future action, distributionally conditioned: "*You will of course not meet him until he has spoken to me.*"

The perfect sets on other axes of orientation have the same meaning as *Pr.P* (in the past and future): "*Dick's safe.*" "*But the soldier said you had been taken.*" "*But if I don't go through with it. where will the heroism be? I shall simply have tricked them ...*".

There is hardly any radical difference in the content of *Pt.P* and *FlP* from *Pr.P*; they are certainly parallel and correspond to the axes they are placed upon. The syntagmatic analysis of tenses (which, however, it not part of our task, limited to the consideration of oppositions), is bound to confirm the invariants of perfect sets established here. These invariant meanings are

7 Cf. Diver (1963) 147.
8 Diver (1963) 151.

also similar in other microsystems of the passive, subjunctive mood, and the verbals.

We have, thus, established the difference of the opposition in perfectness from other types of oppositions in the tense-aspect *MS*, which is verified also by its relation to neutralization. Neutralization of the given opposition, due to its non-privative nature, is also a rare case. But if in the opposition 'present–future' the plane of expression may be neutralized and in the opposition 'present–past' neutralization occurs through transposition, then in rare cases of neutralizing the opposition of simultaneity — anteriority we have a contextually conditioned neutralization of the plane of content of the stronger member of the opposition, i.e. anteriority. Such examples are not numerous: *I thought that she lived in Moscow till she entered the Moscow University*, or on the axis of orientation of the present in set expressions of the type: *I hear you are one of the famous three*, etc.

13.

CORRELATION OF THE MEMBERS OF PRIVATIVE OPPOSITIONS IN ASPECT (DURATION)

Quite a different relation may be seen among members of privative oppositions in the tense-aspect *MS*. To define the nature of the oppositions themselves, we should first establish the invariant of their constituents. It may be said that the two main trends in the treatment of the content invariant of continuous sets are the following: some linguists distinguish the meaning of duration, continuation in the durative set, no matter whether they call it aspect or not; others find in it only simultaneity, i.e. a purely relative meaning. The proponents of the first trend are more numerous, and they point out that the quality of the continuous set is to serve as a frame which limits the process of the action, making it more concrete and isolated. The first point of view is also supported by A. Bodelsen who, however, sees the main difference between non-continuous and continuous sets in that the first state facts, describe usual EVENTS or their result, while the second describe the ACTIONS themselves. The element of duration is thought by A. Bodelsen to be secondary, because in his opinion any ACTION is associated with continuation, whereas the EVENT does not produce any idea of duration. Thus, comparing the examples *I have read Hamlet* and *I have been reading Hamlet*, A. Bodelsen notes that the first states the result of the action, while in the second the action itself is stressed. The same is said concerning the following sentences: *cows eat grass* and *the cow is eating grass*; *he thinks of you–he is thinking of you*; *England wins the test match–England is winning the test match*; in the continuous form the interest is centered on the action itself, and not on the event described.[1]

The other trend, evidently, stems from A. Brusendorf who, admitting that continuous forms express aspect, nevertheless is of the opinion that this aspect expresses a relative notion of simultaneity, and not duration.[2]

Do these two points of view contradict each other, as do the opinions on the invariant of the perfect? In this case, these two understandings of the nature of continuous forms supplement each other. This can be confirmed if we consider the kinds of oppositions in the tense-aspect *MS*.

Proceeding from the binary principle of the organization of the verb system, it is necessary to verify whether the binary framework can cover the sets of continuous and perfect continuous according to their invariant meanings. It is necessary also to elucidate whether or not these sets possess different categorial meanings, as according to many grammarians, or one common invariant meaning.

The oppositions singled out here come into the temporal frame consisting of four divisions in time, each forming an independent axis of orientation in which two divisions of non-perfect–perfect sets are laid within each temporal axis:

	(a) (b)	(c) (d)
present	non-perfect	perfect
past	non-perfect	perfect
future I	non-perfect	perfect
future II	non-perfect	perfect

The sets, coming into the zone of simultaneity (non-perfect) or anteriority (perfect) on different axes of time, should have a common basis for comparison (depending on the temporal landmark), reflected in marked or unmarked *DF*. Thus, e.g., in the non-perfect set on the past axis of orientation, the basis for comparison in the

1 Bodelsen (1936-1937) 220-238.

2 Brusendorf (1930) 225-248.

opposition of the sets (a) and (b) will be one marked *DF* of tense and one unmarked *DF* of simultaneity of the action with the temporal axis. In the perfect set the basis for comparison in the opposition of the sets (c) and (d) will be two marked *DF*, those of tense and temporal relativity.

The correlation of the sets (a) and (b), on the one hand, and (c) and (d) on the other is built upon the privative nature of oppositions, in which the basis for comparison is a more general *DF* of the axis of orientation and a more particular *DF* of temporal relativity within one axis of orientation. This basis for comparison is marked not only in the plane of content, but in morphological formants as well, for example, in forms (1) *write*, (2) *have written* are present: (1) The designator of the present axis plus simultaneity, (2) the designator of the present axis in the form of the auxiliary verb, and anteriority in the combination of *have* with participle II, whereas the contrast is based upon the presence of the aspectual *DF* in the intensive member — *am writing*, *have been writing* (the designator of which is the combination of *be* and participle I) — and upon its absence in the extensive member (sets a) and c)) — *write, have written.*

The privative nature of these oppositions and the isomorphism of their members are confirmed by their base — the meaning of simultaneity in one pair (a) and (b) and anteriority in the other pair (c) and (d); the above features are not repeated in other oppositions, and, hence are natural only for every privative opposition. Thus, the first pair of grammemes, (a) and (b), on all axes of orientation forms an opposition in durative aspect, the invariant content of which, according to its name, is the expression of a durative action in the strong member. This notion covers many definitions of the nature of continuous forms. H. Marchand points out that the progressive form is marked and the simple one is not. These forms cannot be substituted by forms of the present tense in English for they are grammatically relevant, while in Spanish and Italian these two sets of forms are variants of one taxeme and are interchangeable. H. Marchand explains this event by the fact that in English there has never been any contrast of perfective–non-perfective aspect, while in Spanish and Italian these aspects existed.[3]

It would be of great interest for syntagmatic analysis to establish cases when it is possible to substitute continuous forms by simple ones and vice versa, i.e. to establish all concrete conditions of the distribution of tense-aspect grammemes which contribute to the neutralization of opposition in duration–non-duration.

This problem, however, dealing with the syntagmatics of the verb, calls for a description of a mass of linguistic examples, which is impossible within the purely paradigmatic tasks of this manual.[4]

For example, we may compare the following sentences: "*She looks forward to floating your wife in society in London*", where the interchange is possible: *she is looking forward.* The substitution is doubtful in the following example: "*That sounds civilized and comfortable. I vote we go there*", and it is completely excluded in the sentence: "*What do I care about my duty?*".

The continuous form accentuates the very fact of the process of the action at the moment when it is being mentioned; it concentrates the attention on the duration of the process, which makes it impossible to use it in narration for setting forth the chronology of separate events. It is the coming of the continuous set into the relative temporal frames of either simultaneity or anteriority, that creates the concreteness and narrowness of the action, which has been noticed by many linguists. This also leads to isomorphism of the content of the members of the opposition in duration, thus enabling the marked member to be neutralized and substituted by the unmarked one possessing a more general meaning which includes both members. The fact of both members of privative oppositions belonging to the same relative temporal plane makes their interchangeability possible where no definite restrictions are imposed by the invariant meaning of the members of the opposition, or by their distribution. If we compare the two

[3] Marchand (1955) 45-52.

[4] The following articles present a considerable interest in the sphere of syntagmatic morphology of the English tenses: Diver (1963); Calver (1946) 317-325; Hatcher (1951).

sentences *she is just coming out, I think, her carriage is waiting for her* and *she comes out every day at the same hour*, we shall see that the goal of the first sentence is to express the process itself at the moment of speech, while the second shows habitual action with its unlimited nature, which excludes the use of continuous forms. In such contextual conditions the neutralization of the given opposition is impossible. A case of neutralizing *Pr.C* is in the example given by W. Diver with the description of the process of playing ball: "*the pitcher takes his stretch. He looks at first where Mays is taking a good lead*",[5] where both members of the privative opposition appear in turn, though the nature of the action is similar. We also have neutralization in the example "*I tell you, Nora, I'm going to develop this place.*" The neutralization of the relevance of duration is more usual on the axis of the past, following from the narrative function and interconnection of the past tenses, like in the sentences *I was sitting (or sat) while she was standing (stood).*

The invariant meaning of the continuous set also explains the limitations of its use, mainly lexical ones, for example with verbs of indefinite duration as *like, remember, love,* since duration, expressed by the *C* form, is always limited and is natural for isolated actions in the stage of development.[6] The stylistic use of the *C* set, as usual, is in agreement with its invariant when duration is stressed on purpose, as is the case in the sentences like *I am always expecting something; she is always refusing people.*

In languages which lack the continuous form, actions of different duration and range are successfully expressed in one common form within one relative temporal plane. It is the privative nature of oppositions in aspect (in English) that testifies to the fact that aspectual forms with the *DF* of duration are easily substituted.

They may even be lacking altogether in systems built on the relative temporal division of simultaneity and anteriority (e.g., Scandinavian languages). Hence, the durative set is a source of redundancy; its grammemes on temporal axes, other than those of

[5] Diver (1963) 174.
[6] Marchand (1955) 45-52.

present-past, can be easily omitted in speech. The investigation of the possibility of neutralization within oppositions in aspect, as well as close scrutiny of conditions of context where, by the norms of the language, either one or the other member of the opposition should be present, yields much more interesting results than the description of the use of separate formtypes, as is usually the case in manuals on tenses. It is the oppositional analysis that reveals new perspectives of the study and description of concrete uses of forms in speech.

A detailed analysis of the correlation of the lexical meaning of verbs and grammemes, of conditions of the context, etc. is contained in numerous theses and books on tenses, although this analysis would have been more useful on the basis of *DF* of grammemes, their oppositions, and the invariant grammatical content penetrating all cases of use and interacting with the data of inner (lexical) and outer (syntactic) distribution. Such an analysis would help to give reasons for every case of the use of tenses and define the rules and deviations from them. The invariant of the meaning of *C* sets may be discovered even in a few examples: "*Are you coming Dr. Trench?*" "*I am doing my duty, I am clinging to my duty! My duty is to get you away...*" "*... nonsense is always insincere*; and *my dearest is talking nonsense.*" "*Trench and I were putting our heads together over the letter just now.*" "*When I interrupted your conversation here some time ago, you and she were evidently regarding it as settled.*" "*We were thinking of going on to Rolandseck.*" "*I thought you were detaining me because you had something to say.*" "*How much more manly it would have been to confess that you were trifling with me that time on the Rhine!*" "*I am afraid he will be thinking more of his own danger than of mine.*"

All the *C* forms accentuate a definite, durative nature of the action in the plane of simultaneity with the given temporal axis; they correlate with the initial, unmarked member of the opposition — the indefinite form. The temporal plane of the members of the opposition is common and is equally revealed in all the four vertical sets I, II, III, IV shown on page 112. This is an opposition whose identical *DF* are the marker of the temporal plane and simultaneity with it, and the element of distinction is the presence or absence of aspectual characteristics. Hence, one cannot agree with the view that simultaneity is the basic meaning of *C* forms, since it is but the temporal side of the meaning of these forms, their basis for comparison, while duration is an aspectual seme, constituting the main content of the strong member of the privative opposition. Aspect in English is subordinate to tense (direct or relative), and aspectual grammemes cannot have a purely aspectual meaning outside the temporal one. That is why all aspectual forms are relative. In regard to the weak member of aspectual oppositions, the continuous form always shows simultaneity. Sometimes the temporal meaning of the form, depending on the contextual conditions, may be advanced to forefront, shading out the aspectual meaning, but that occurs mostly in the zone of the past. The relative content of the form consists precisely in the coexistence of the action it marks with other actions of the same temporal plane. That is why, in setting forth a chain of actions of a different nature but taking place at the same time, the sets of indefinite and continuous may alternate, especially in the plane of the past, since the isolation of the forms of the past and their narrative nature creates highly favourable conditions for limiting the action expressed by *Pt.C* to the sphere of another action on the same axis of the past, in the process of narrating actions and events of the past. The length or shortness of the action is not accentuated in the indefinite set, since it is neutral to aspect; duration is singled out in special aspectual forms. The length of the action denoted in continuous sets is always limited and cannot be infinite, since the process, marked by the *C* set, is bound by relative frames of simultaneity with that temporal plane within which set *C* exists. The specific categorial destination of three grammemes of the axis of the past may be traced in the following example: "*On Friday night about eleven he had packed his bag, and was leaning out of the window half miserable, and half lost in a dream of Paddington Station, when he heard a tiny sound, as of a finger-nail tapping the door. He rushed to it and listened.*" (J. Galsworthy).

The peculiarity of the temporal axis affects the nature of the use of *C* sets. Thus, simultaneity with the moment of speech implies the meaning of the action taking place at the given moment, while simultaneity with the axis of the past allows for alternation of forms with grammatically expressed duration or without it, when successive events are described.

What is the correlation between sets (c), perfect, and (d), perfect-continuous *(I have [had, will have, would have] written–I have been writing)?* As it should be expected, this correlation is parallel to the correlation between the sets (a) and (b), i.e. it contrasts a durative action with an action neutral to duration. It means that the perfect-continuous *(PC)* forms also express the durative aspect, but only within the relative temporal plane of ANTERIORITY to the given axis of orientation. The sets (c) and (d) do not exclude each other, being isomorphous as are the sets (a) and (b), and that means that here as well, the member marked in duration, can be neutralized, and in position of neutralization an unmarked extensive member with a most general meaning will appear. The basis for comparison in the opposition of the perfect — perfect-continuous sets is the relative stage of anteriority, it being correspondingly marked formally in the combination of *have* + participle II, on which is placed the marker of duration (*be* with participle I).

Comparing continuous forms in the non-perfect and perfect sets, one may be sure that the difference between them is built on a distinction in the nature of temporal reference, with the main categorial content of duration being common for both. The invariant meaning of *PC* forms is the same meaning which distinguishes members of the indefinite-continuous opposition, i.e. the aspectual meaning of duration, similar to that contained in the *C* set. Correspondingly, the semes of the *C* set include a marked (both formally and in content) *DF* of duration (strong seme) and semes showing the basis for comparison — that of the temporal axis, formally expressed (strong seme), and simultaneity, formally not marked (weak seme), while the semes of the *PC* set have THREE STRONG SEMES — duration, perfectness, and the temporal plane of the action (present, past, etc.). The distribution of the semes in different sets of the microsystem corresponds to the place a particular set holds in the oppositions, whether it is a marked or unmarked member of the opposition and of which opposition, privative or equipollent.

The dual nature of the meaning of *PC* forms in the present tense, that is their anteriority to the moment of speech and, at the same time, their connection with it, on the one hand, and on the other, the aspectual *DF* of duration combined with perfectness, fully explains both the inclusive and exclusive meaning of the *PC* type. The combination of the meaning of anterior present and duration explains the fact that *Pr.PC* may express an action which was going on in the anterior present and comes close to the moment of speech, but ends before it (exclusive meaning): *I have been watching you, you look awfully bad-tempered* and it may express an action which continues up to the moment of speech and including it (inclusive meaning): *The tea has been boiling over since half past eleven.* Both inclusive and exclusive use of *Pr.PC* illustrates its invariant content of duration in the sphere of anteriority: "*They have been waiting your convenience for fully half an hour.*" "*Oh, do you think I have not been telling myself that all night?*" "*Well, I have been thinking ever since.*" "*I blush for you. I was never so ashamed in my life. You have been taking advantage of that unprotected girl.*" "*He goes into the hotel; and Cockane, who had been hanging about inquisitively, emerges from the shrubbery.*" "*They said she'd been asking for a separate room.*" (J. Galsworthy).

To sum up the discussion on the invariant meaning of both perfect and non-perfect *C* sets, let us draw attention, once again, to the fact that both sets possess the same leading marked *DF* — that of duration. This means that their aspectual invariant is the same. The difference between the sets lies in their basis for comparison, which includes a different number of strong and weak semes in different correlation with one another.

14.

GENERAL CORRELATION OF OPPOSITIONS AND MORPHOSEMES IN THE TENSE-ASPECT MICROSYSTEM. RECOMMENDATIONS FOR SYNTAGMATIC ANALYSIS

As is clear from the above, the tense-aspect *MS* of the English verb presents two types of contrasts in temporal and aspectual *DF*. Temporal correlation shows itself in two qualities — purely temporal, based on marking temporal axes of orientation in verbal forms, and a relative temporal contrast, based on comparing forms within one temporal plane in the line of simultaneity or anteriority to this temporal plane. Aspectual differentiation takes place in the opposition of a durative action to an action devoid of any duration mark, within both simultaneity and anteriority to the given temporal plane.

Every opposition consists of two members, one of which is the initial member (no matter what is the nature of the opposition), which does not mark the contrasted *DF*, is neutral to it, and can have a zero marker of the *DF* (in a privative opposition), while the second member necessarily marks the contrasted *DF* and has a special morphological formant for it.

In sets I and II oppositions in tense and aspect are realized most fully and logically, due to the vitality of these two temporal planes, their constant repetition in the process of communication, and their genetically-caused dichotomy. Aspectual and relative temporal differentiation in sets III and IV is insufficiently realized due to the rarity of the situations created by these formants. A future projected action is usually considered as a whole, as a fact, and the speaker can very well do without mentally defining its possible duration or anteriority. The farther right of the indefinite set the set is situated, the more seldom it is used, and, finally, *PC* on the axes of future *(I shall (should) have been writing)* is actually never used in narration and may find application in personal speech for the creation of a certain stylistic effect or for the deliberate stressing of some action. However, these seldom-used verbal units cannot be excluded from the system of forms, since they exist not in one's imagination, but in the language, as potential members of linguistic oppositions, though they do represent the redundancy of the system.

All the members of the tense-aspect paradigmatic scheme are interlinked and interdependent; they form a closed circle as a unified whole built, however, on the basis of contrasts. The oppositions within the *MS* intersect by way of expressing identical and contrasted *DF*. This may be shown in the scheme (p. 157).
According to the scheme (1) all oppositions consist of two points, i.e. they are binary, (2) each point (grammeme) contains as many intersections as it has *DF* (i.e. three) on which the oppositions are based, (3) the most dense cluster of initial members of oppositions is the grammeme *Pr.I*, (4) the clusters of oppositions become more sparse as they descend along the vertical line (from present to future), (5) oppositions at all layers of the system are proportional, (6) temporal oppositions (including relative ones) prevail over aspectual ones and penetrate all corners of the system, (7) every two points linked between themselves are in opposition and are correspondingly called 'intersecting'. For example: *Pt.PC* is in opposition to *FIIPC*; *Pr.C* to *Pt.C*, etc. In accordance with the members of oppositions, the bases for comparison are also changed. Thus, for example, the basis for comparing *Pr.C* and *Pt.C* is the seme of duration, and the temporal *DF* represents the contrast.

The content of the oppositions in which the forms participate clearly affects their morphological structure. Every strong *DF* of the intensive member finds formal expression in a grammatical exponent. For example, *Pt. PC* — *had been speaking* — contains the combination of *be* with participle I, a marker of durative aspect

SCHEME

of intersecting binary oppositions in the tense-aspect microsystem (active)

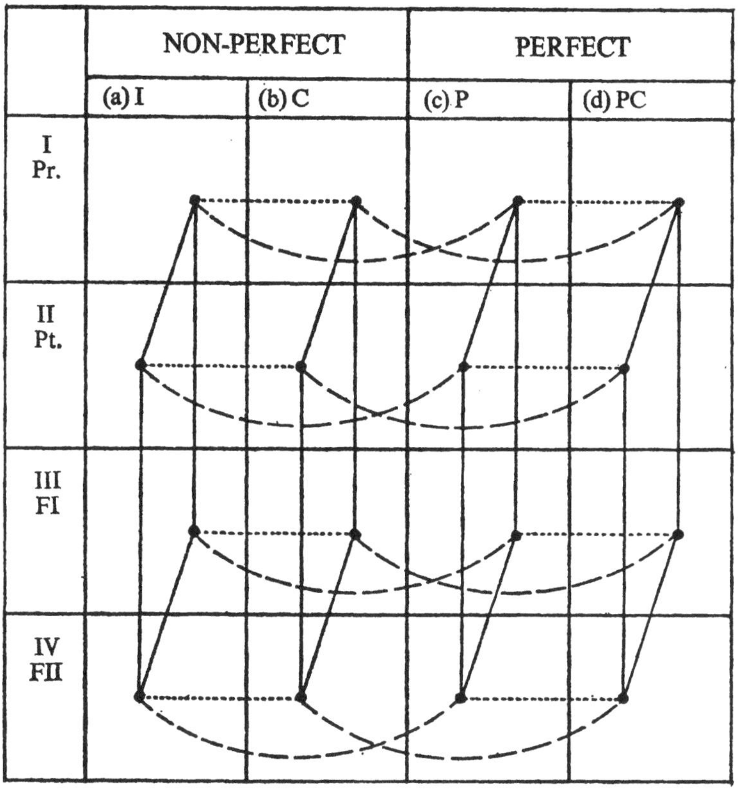

The straight line shows oppositions in tense.
The dotted line shows oppositions in aspect (duration).
The broken line shows oppositions in temporal relativity (perfectness).

plus the form of the present tense of the finite verb, informing of the axis of orientation on which the grammeme *Pt.PC* is placed; the combination of *have* + participle II signals the *DF* of perfectness. The extensive member may have zero grammatical markers of the relevant *DF* of the opposition, but it always contains the basis for comparing the members of the opposition. Thus, *Pr.I* signals about the axis of orientation of the action and about the reference to the moment of speech, which serves as a basis for comparison of both temporal oppositions and oppositions in temporal relativity, and also in aspect on the present axis of orientation. The aspectual *DF* is not marked in this grammeme, either formally or in content.

In view of the above considerations, it would be hardly expedient to ascribe to each separate form the meaning of a special aspect or tense, as is the case in a number of works and theses devoted to the analysis of some formtype, without taking into account the recurrence in every layer of the microsystem of analogous relations which manifest themselves in binary oppositions based on identical and contrasted *DF* of verbal categories. An atomistic analysis of separate units cut off from the system, without considering the relations they enter into with other members of the system, cannot produce an objective evaluation of the nature of grammatical categories. A definite given correlation not only passes consecutively through homogeneous layers of the microsystem, but is also reflected in other microsystems of verbal forms in units similar in their morphological structure. Thus, for example, the invariant meaning of the combination of *have* with participle II is retained in other parts of the verbal system as well; such is the natural logic of the structure and the principles it is built upon. In other words, perfect sets in the non-finites, in the subjunctive mood, and in the microsystem of voice have as their invariant the same content as is contained in the same sets of the tense-aspect *MS*. The continuous forms, accordingly, have an analagous categorial meaning in all the small systems of verbal units in which they appear.

No formtype denotes grammatical categories outside the opposition in identical and contrasted *DF* without repeating this contrast in other sets of homogeneous formtypes. This is a sign of the pressure of the system. What units of the tense-aspect *MS* may after all be in opposition? The answer stems from the very nature of opposition in general, which, though implying contrast, always has a basis for comparison — some *DF* common for both members, since otherwise there can be no grammatical opposition.

Both the common and contrasting features are those *DF* which enter the invariant content of the morphological unit as separate semes; however, every grammeme may have stronger and weaker semes. Strong semes are always marked, since they form the basis for distinction, while the weak semes are unmarked.

Hence, members of one set may find themselves in opposition, either along the vertical or horizontal line of the *MS*, since every set is built on the basis of a common *DF* making up the basis for comparison. Thus, grammemes of the continuous sets of different temporal planes or grammemes of the past set (plane) may be opposed to one another. But here as well, not all the members of every set may be contrasted in any order, since a more general *DF* of the whole set (present, for instance) is overlain by another *DF* inside the set (say, anteriority), which limits the compatibility of the members of the whole temporal plane in one opposition. So, on each temporal axis the second relevant *DF* will be simultaneity or anteriority, and only those members of the given temporal axis will be contrasted which are linked by one of these general *DF* common for both members. Hence, the indefinite set is not contrasted to the *PC* one, and the continuous set is not contrasted to the perfect, since they have different bases for comparison and more than one *DF* for distinction, something which is impossible for an opposition.

It would be worth mentioning other points of view on the nature of oppositions in the English tense-aspect *MS*. Thus, e.g., J. Kurylowicz in his article, "Aspect et temps dans l'histoire du persan"[1] considers the contrast of imperfective and perfective aspects as the basic one in any verbal system, though he points out that in Western languages the notion of perfectivity gives way to the notion of anteriority.

Proceeding from an aspectual understanding of the basis of any verbal system the following scheme is suggested: the contrast of the imperfective (negative member) and perfective (positive one) implies the existence of a neutral form (neither imperfective,

1 Kurylowicz (1953) 109-118.

nor perfective), and a complex form (combining elements of both the perfective and imperfective). With such a division, the English indefinite set falls under the heading of the neutral member, the durative set under the imperfective member, the perfect set under the heading of the perfective one, and *PC* is a complex form. This series of contrasts is admitted only on the axis of the present and past, while future is thought to be a mood and excluded from temporal divisions. However, much in this system remains vague. If the indefinite is a neutral member, then it is not clear to what opposition it belongs. We may contrast two elements, otherwise there is no opposition. On the other hand, if the *C* set is an imperfective, negative member, why then is the imperfectiveness in this negative member so clearly marked? Great doubt arises about the notion of a complex member, combining in itself two aspects opposite in content. It is difficult to imagine a language in which the perfective and imperfective aspects would be marked at the same time in one form. The contrasts in European languages are evidently based on a different principle, and their frame is expressed not by any aspect, but by temporal relations, about which speaks J. Kurylowicz himself. In J. Kurylowicz's scheme the basis for comparison of oppositions remains undefined. These judgements lead us to the question: are not the sets *C* and *P* really contrasted in the English tense-aspect *MS (I am writing–I have written)?* If the system possessed these two formtypes only, it would be possible to speak of their contrast to each other in an absolutely equipollent opposition with equal degree of markedness of the aspectual *DF* in different morphemes and with different content. Such a situation, evidently, is not natural for any opposition generally, since an absolute equality of its members (the absence of a certain hierarchy) suspends their contrast. Besides, should this opposition be accepted, for the English verb, other sets would remain completely outside oppositions in the system, and that is hardly possible for a system which shows the recurrence of equal relations in different sets.

It follows from the above that the oppositions in this case are built on a different basis, which we tried to discover in this work.

A serious factor in determining the nature of oppositions is also its formal markers. The different structural composition of the formal elements of *C* and *P* sets testifies to their belonging to different oppositions.

Belonging to different oppositions, however, does not exclude contiguity among their members, an example of which are the forms of *Pt.I* and *Pr.P* (*wrote* and *has written*). Much depends upon the concrete conditions of the temporal reference and the specific nature of temporal relations, as well as on the point of view from which the action is considered.

Differentiation of categorial filling of different verbal grammemes interacts, besides other factors, with the need to distinguish linear actions possessing a certain length, and momentary, point actions which Brunot calls '*actions ligne*' and '*actions point*';[2] various categorial forms distinguish these kinds of actions in a different way. Thus, in members of oppositions marked in aspect and temporal relativity, linearity prevails in durative forms and point actions prevail in the perfect, the type of action defining the invariant categorial content of the forms and, in its turn, being defined by it. From here is the degree of capability of the form to participate in the narration or to convey an action which is limited by a definite length in the zone of simultaneity. The unmarked members reveal to a lesser degree the type of action and can, therefore, be more mobile and meet the varied needs of communication — to convey a narration, etc.

The categorial content of grammemes in the English tense-aspect is considered by analysts from different angles; a detailed treatment of all avenues of approach would take too much space. We may only note that those linguists who ascribe aspect (terminative, perfective, transmissive, etc.) to the perfect set (e.g. I. Ivanova)[3] fail to give a definition of the categorial content contained in the perfect-continuous set, since the two aspects (durative and perfective) cannot naturally combine in one grammeme. With such a definition of the basic grammatical meaning

[2] Brunot (1953) 438.
[3] Ivanova (1961) 147-149.

of the English perfect is also linked the postulation of anteriority only within the past perfect and future perfect sets (in *Pr.P* it is not recognized). Besides, ascribing an aspect to the perfect leaves beyond the system the categorial content of perfect sets in other microsystems — non-finites, subjunctive, etc. It is, however, difficult to suppose that transmissive or perfective aspect must be marked in these microsystems as well in the form of have+-ed.

In such schemes of the categorial content of the tense-aspect paradigm the question remains open, which sets are contrasted to each other and in what *DF*, and the proportion of oppositions is violated, although it is natural for any complex morphological system, such as the verb.

More rational are such schemes of tenses where the proportional nature of sets is observed. Thus, A. Smirnickij postulates in the indefinite sets the content of common aspect, temporal reference, and tense; in the *C* and *PC* sets, the content of aspect (durative); and in the perfect sets, the temporal reference (anteriority). However, the nature and direction of oppositions and the *DF* of the sets are not mentioned by the author. In addition, the scheme of A. Smirnickij completely excludes the axis of orientation of the future action considered from the point of view of the past (*F* II), though this axis is absolutely parallel to the axis of the simple future (cf. *I think he will agree* = *I thought he would agree*).[4] Different modifications of such a scheme cause other irrationalities: the scheme either completely excludes the category of tense (time) (together with *F* II).[5] which, however, is quite clearly marked both by form and by meaning in all the sets of the microsystem, or absolutization of the temporal reference is taking place and the continuous set is thought to be an exponent of simultaneity.[6] In this case the strong member of oppositions marked in special exponents of *be+-ing*, loses its own categorial meaning, simultaneity, being however, only the basis for comparison of two members of the opposition.

4 Smirnickij (1956) 136 ff.
5 As with Barkhudarov and Steling (1960) 202.
6 Irtenjeva (1956) 77.

All these short remarks testify to the fact that in defining the categorial meaning of grammemes and, mainly, their place in the system, there exist considerable difficulties to overcome which, as we think, is possible only by means of identification of the oppositional correlations within verbal sets in microsystems and in the whole macrosystem of the verb.

To conclude the analysis of the microsystem of tense-aspect, it would be expedient to give a definition valid for English of those grammatical categories which find their formal expression in it. These categories are three in number: tense, temporal relativity (perfectness), and aspect (durative) (about the category of person — number see chapter 4).

TENSE, in the opposition of grammemes of the indefinite set to each other, expresses the relation of the action denoted by the verb to the present moment (act of speech). This relation manifests itself first of all in the opposition of grammemes of *Pt.I* showing the coincidence with the act of speech and *Pt.I* expressing anteriority to it. This is the main dichotomy of the category of tense in the English verb, formally marked in the base of the verb (weaker member of the opposition) and in the markers modifying this base (in the past — the stronger member of the opposition). Furthermore, tense denotes the relation of posteriority to the moment of narration in the present or past, depending on which the future action is marked either by formants including elements of the present or of the past. The formal dichotomy of present-past is thus preserved in the means of denoting actions following the moment of speech as well, i.e. future ones. Posteriority to the moment of narration is expressed in the indefinite grammemes of the axes of *F* I and *F* II.

Thus, tense in the English tense-aspect *MS* expresses the relation of an action to the present moment, on the one hand, and to the moment of narration in the present or past, on the other.

ASPECT expresses in the sets *C* and *PC* (marked members) a durative action limited by the frames of simultaneity or anteriority to the given axis of orientation.

Such a definition of aspect in English, we think, is of great theoretical and practical importance, since it removes the basis for radical differences in the opinions of linguists on this problem. Solving the question of the English aspect on the basis of the opposition of marked–unmarked member with a common basis for comparison (the opposition being verified in the process of neutralization) removes: (a) the notion of common aspect, (b) the possibility of looking for contrasts of the Slavic perfective–imperfective aspects in the English verb, and (c) the contradiction in treating the perfect, since the existence in the system of one type of opposition (e.g. in duration) implies also the presence of another type (in simultaneity–anteriority).

Such an interpretation is also justified historically: continuous grammemes appeared within the frame of relative temporal divisions, and that is why these forms are becoming a source of redundancy in the verbal macrosystem.[7]

TEMPORAL RELATIVITY expresses an action simultaneous (in the indefinite set, the weaker member of the opposition) and anterior (in the perfect set, the stronger member of the opposition) to the point of reference of the given axis of orientation. The point of reference is represented by the temporal axis marked in indefinite grammemes. The difference between tense and temporal relativity consists not only in every temporal form denoting a new temporal axis of orientation, but also in a different nature of the intervals marked by the above categories, during which the action takes place. Thus, the interval, during which an action may continue, marked by the *Pr.P*, is limited by the frames of the present and cannot go beyond it, since then it comes across another axis of orientation, first of all, the past, which limits its 'movement' to

[7] On the use of this term, mathematical in origin, see Spang-Hanssen (1964) 61-66. On redundancy of phonological *DF* see Jakobson and others (1955) 6; Malmberg (1963) 113. However, redundancy of phonological and morphological units are different phenomena due to the bilateral nature of meaningful units. Cf. Bull (1960) 32. Redundancy is connected with the pressure of the system ((*Systemzwang*), because the appearance of any new form in the system brings about the appearance of a similar form (or forms) in parallel sets or microsystems (Cf. the sphere of the development of English continuous forms in parallel sets). On the pressure of the system and different aspects of this notion see Martinet (1960) chapt. 6, § IV.

the left, i.e. into the past. Future perfect is limited by the axis of the present. That is why the above grammemes are not used in narration, for they are bound by the direction of the action (its vector) towards the point of reference, and by the interval between the two temporal axes.

The direction of the action denoted by the past perfect has less strict boundaries, since, although directed towards the point of reference (the axis of the past) the past perfect has no barrier in the way of a preceding axis of orientation, since the temporal divisions are devoid of an axis which would be capable of preceding the past. This explains the ability of *Pt.P* to take part in narration.

Double limitation in temporal relation is a feature of the grammemes expressing the durative aspect, since their limitation goes both along the line of simultaneity and anteriority to the given temporal plane, and in addition they are covered by the limit of the sphere of action on different axes of orientation.

The forms of the indefinite, each denoting a special axis of tense, are infinite, unlimited by any intervals. This set, therefore, is used in narration to convey general, universal, and habitual actions.

The invariant of every set established so far is based upon

	Simultaneity		Anteriority	
time	Indefinite	Continuous	Perfect	Perfect Cont.
Grammatical categories *(DF)*				
Pr.	tense	durative aspect	temporal relativity	durative aspect
Pt.	tense	durative aspect	temporal relativity	durative aspect
F I	tense	durative aspect	temporal relativity	durative aspect
F II	tense	durative aspect	temporal relativity	durative aspect

correlations which actually define the limit and sphere of the realization of the categorial meaning of any grammeme.

The above system of categories may be set forth in a table which shows the BASIC categorial (marked) *DF* of the set (no matter whether the grammatical category is realized in a privative, equipollent or an intermediate kind of opposition). (page 165). According to the table and our idea of the categorial filling of verbal grammemes, the microsystem turns out to be symmetrical and meets the characteristics of an ideal system of tense, distinguished by its symmetry from systems of other European languages. Presented in the above way it differs from the systems built by grammarians, mentioned (and not mentioned) above.

Since our analysis of tense-aspect is mainly limited within the frames of paradigmatics, with only a few elements of syntagmatics (linguistic examples) used to illustrate some postulates, we may only map out the tasks of the syntagmatic analysis of the verb system which follow from the analysis of oppositions.

The advantage of a syntagmatic analysis on the basis of previously defined oppositions seems to be beyond a question, due to a number of reasons. For practical purposes it is important not only to give a list of different meanings and instances of the use of each separate form (discarding their mutual link), which is rather difficult to comprehend, but also to produce a logically motivated distinction of one member of an opposition from another, on which their distribution in speech is based. Thus, for example, research would be valuable which would show in which cases only the indefinite set is possible and not the continuous one, in which cases duration must be expressed and when both sets are interchangeable, when the meaning of the strong member may be neutralized, what lexical and distributional limitations are imposed upon both members of the opposition in aspect, etc. — all this being determined in terms of oppositions and on the basis of the invariant of the grammatical meaning of the grammemes. Contiguity between the members of different oppositions (as, e.g., between *Pr.P* and *Pt.I*) may also be more rationally studied in the plane of the oppositions into which each of them enters — it would help to clear up the reason for the distinction and contiguity between them. Oppositions and the invariant also supply the explanation for the variant meanings conditioned by the context, as well as for those arising on the basis of special processes, such as neutralization or transposition. Thus, it is difficult to understand why *Pr.P* may acquire the meaning of future; under no conditions can this meaning enter its invariant; however, the analysis of oppositions and cases of their neutralization reveals the source of such syntagmatically conditioned usage. It would also be useful to check the action of the morphosemes, as well as which lexical subclasses of verbs possess all the paradigmatic forms, and which ones have some paradigmatic limitations (for instance, the use of *C* sets with verbs denoting feeling, etc.).

For the purpose of such an analysis it would be fruitful to use the method of substitution (which helps reveal the relevant *DF* of the form), and also transposition into another temporal plane.

The parallelism of the meaning of analagous grammemes in various sets is verified on the basis of oppositions and their proportional nature. For example, the parallelism in the meaning of *Pr.P* and *Pt.P* may be verified by a transposition of the narration taking place in the past into the sphere of the present. In this case we can verify whether the actions preceding the moment of narration in the past which are marked in *Pt.P* would really be properly conveyed by the *Pr.P* in a narration taking place at the moment of speech, in the present.

It is natural that the revealing of units of the immediate environment, relevant for the realization in connected speech of the grammatical meaning of the form, is also of great importance. For example, the qualities of the members of an opposition in temporal relativity may be verified by their ability to combine with the article. For example, the definite article cannot be used in such sentences as *I have received the new pen as a gift.*[8] Of great significance is the study of lexical distribution, also in the plane of oppositions, their qualities, and the qualities of the axes of

[8] Diver (1963) 157.

orientation on which they act. Thus, according to the norms of the language, the following sentences are impossible: *She has cried or she has played cards, I am loving you, my nose runs,*[9] etc. Of great interest is the analysis of syntagmatic conditions of neutralization of oppositions.

Thus, the next stage in describing the verbal system will be the testing of oppositions in action, their proportionality, the possibility of interchange of members of one opposition and different ones, and the study of their interaction with data of the inner (lexical) and outer (syntagmatic) distribution.

Although the syntagmatic analysis of the use of verbal forms in speech (which we refer to the field of syntagmatic morphology) is not part of this publication, to support the above theory we shall cite some examples of verbal forms *(VF)* in context, pointing out the ways of their interpretation in the light of the theory of oppositions. But before this it is necessary to remind the reader of the following: THE PARADIGMATICS of *VF* is the establishing of the invariant, systemic (categorial) meaning, of *VF* as members of a paradigmatic set, on the basis of associative ties among grammemes united by the sum of formal and content features into one paradigmatic microsystem. Any *MS* consists of a network of oppositions of the meanings of *VF*; the composition of oppositions and their direction depend on *DF* making up the invariant of the grammatical meaning of *VF*. Paradigmatics is an event not to be directly observed, but deduced on the basis of definite abstract operations.

The SYNTAGMATICS of *VF* is the interaction of the invariant meaning of *VF* and linguistic means conditioning the realization or neutralization of the invariant meaning *(IM)* of members of verbal oppositions in speech, i.e. in a linear chain; the event is directly observable.

If establishing the paradigmatics of *VF* takes place through the method of oppositions, the testing of the action of oppositions on the syntagmatic axis is realized mainly by the method of distribution, taking into account the types and units of the context.

[9] Hatcher (1951) 254.

It should be expected that in the sentence either a realization of the invariant *(IM)* or its neutralization may take place. If the context lacks units causing the neutralization of some opposition, then the *IM* of one of the members of the opposition is usually realized, i.e. the paradigmatic meaning coincides with the syntagmatic one. In the sentences: *Do you smoke? We shall meet again. He died abroad. He hoped he would enjoy the show*, *IM* of the indefinite sets is realized. In the sentences: "*I am not disputing the point, I am looking at it from the outside point of view.*" (A. Christie). "*What were you drinking, may I ask!*" (B. Shaw). "*They will be calling each other sister.*" (O. Wilde), we have the realization of *IM* of the continuous sets. In regard to *Pt.C* it may be said that its grammatical meaning is more often realized in a wider context possessing a number of actions succeeding one another and in which the background is the form of *Pt.I.* Here are examples of the realization of *IM* of the perfect set, with an action immediately preceding the temporal axis: "*You've eaten at least a dozen figs.*" (J. Joyce). "*The Yankees had come, the darkees had gone out and Mother had died*". (M. Mitchell); the perfect-continuous: "*She has been seeing a lot of him*". (A. Christie). "*I perceived that she had been crying.*" (J. Joyce).

For the past and future perfect, as well as for the *PC* forms, the context of a complex sentence is more natural.

In syntagmatic analysis one should also consider the statistical potential of forms, i.e. the presence of redundant units, such as *FIIPC*, *FIC*, *FIP*, etc., which are rarely to be met on the syntagmatic axis.

All the above examples show a neutral context, without contextually active units, which would support *IM* of *VF* and serve as 'realizers' of this meaning. *VF*, however, may also appear in an active context possessing realizers of *IM* of *VF*. Such realizers are usually represented by various adverbial modifiers. Drawing up an inventory of the realizers of *IM* of *VF* is certainly an essential task of syntagmatic analysis. For example: "*I snore at night sometimes.*" (M. Twain). "*From time to time, as she took her slow stitches, she raised her eyes.*" (J. Joyce). Here the adverbial modifiers are realizers of an iterative action which comes into *IM* of forms of the indefinite set. Substitution of *VF* in these sentences by the countermember of the opposition *(was snoring, was raising)* would give ungrammatical sentences.

The function of realizers of *IM* may be performed by different adverbial modifiers of the following type: *In the afternoon, in July, the next morning, just now, two years ago, still, for half an hour, long*, etc., different adverbial modifiers being capable of supporting *IM* of different sets of *VF*: "*You're still breathing like a quarter horse that's been run a full mile.*" (A. Christie). "*I have for a month past been living the life of a mountain goat.*" (B. Shaw).

However, in the context (immediate environment of *VF*), units are found which cause neutralization of the opposition, when *IM* of the extensive member appearing in position of neutralization is widened, and the meaning of the intensive, marked member becomes completely neutralized, in other words, the *DF*, which is a sign of the distinction of one member of the opposition from the other, is neutralized.

'Neutralizers' may be expressed by adverbial modifiers, the lexical meaning of the verb, a certain pattern of the sentence, or the general situation of the utterance. For example, in sentences: *I adore you, I see the big idea now*, we have cases of constant neutralization caused by the lexical meaning of the verb. In the next sentence, neutralization of future is due to a definite type of sentence: "*You will of course not meet me until he has spoken to me.*" In the sentence "*I have known him over a period of some years*", there are two neutralizers of *Pr.PC* — the meaning of the verb and the adverbial modifier. The content of duration, contained within the adverbial modifier, makes the use of *Pr.PC* superfluous. The adverbial modifiers in the following sentences contribute to the neutralization of the future: "*My father is coming here to-night.*" "*How long do you stay here?*"

The following are examples of neutralization of the opposition *Pt.I — Pt.C*: "*I cried all night after doing that.*" (B. Shaw); "*They looked at each other for a long moment*" (A. Cronin). In the sentence: "*By 12 o'clock, with the sun pouring into the room the heat became quite oppressive*" (J. K. Jerome), we have the neutralization of the perfect.

In all the above examples of neutralization, the substitution of a counter-member of the neutralized opposition is possible (e.g. *I was crying all night*; *How long will you stay here*, etc.).

Such, briefly, is the suggested trend of the syntagmatic analysis of the use of tense-aspect *VF*, which follows from the above theory of oppositions.

Concluding the analysis of oppositions in the tense aspect *MS* indicative active, which was intended to illustrate the nature, purpose, and also the results of the oppositional method of description of morphological units, their grammatical meaning (categories), it would be expedient to give a general table of combinations of morphosemes throughout the English verbal macrosystem. The present work did not set the goal of giving a broad analysis of all oppositions acting in the whole system of the verb, since it would extend the volume of the given publication. Since, however, the tense-aspect is the initial member of interparadigmatic oppositions and reflects the formal *DF* possible in all verbal microsystems, as well as grammatical categories coming into other microsystems, it is desirable to show the composition of morphosemes (categorial *DF*) in different microsystems.

Table of strong semes distributed in microsystems

Morphosemes (*DF*)	Microsystems			
	tense-asp. active	tense-asp. passive	subj. mood	verbals
tense	+	+	—	—
temporal relativity (perfectness)	+	+	+	+
duration	+	+	+	(+)
unreality	—	—	+	—
passivity	—	+	+	+

The table shows that the universal *DF* is that of perfectness. The plus in brackets under verbals implies an incomplete coverage of the whole microsystem by the seme of duration.

On the basis of these semes are built both intraparadigmatic and interparadigmatic oppositions. The fact that it is these semes that are relevant for the invariant of grammemes may be confirmed by an attempt to synthesize the verbal form on the basis of combining the semes. Thus, if we know that the seme of the past tense has a standard exponent in *-ed*, the seme of duration — in combination of *be* with participle I, etc. — then, knowing the order of distribution of the exponents of the semes we may get, through their different combination, the existing form of a definite verbal unit: the combination of semes of perfectness, passivity, and past produces the grammeme *had been spoken*.

As is clear from the table, the following number of strong morphosemes correlate in the microsystems: in tense-aspect *MS* active — 3; passive — 4 (one marked *DF* of passivity added); in subjunctive (unreal) mood one more *DF* of unreality[10] is added but tense is excluded; the *MS* of verbals has two universal *DF* and one (duration) which does not cover all the non-finite forms.

[10] Many controversial problems and difficulties are connected with the definition of the meaning and even forms (exponents) of the unreal mood in English. Even to mention them, let alone to dwell upon them, would be quite impossible here. All this is dealt with in the book by the present author Khlebnikova (1971).

15.

SUMMARY

Let us enumerate the questions discussed here and draw some conclusions following from the above said.

— The detection of oppositions in morphology is a means of formalizing grammatical semantics, that is, of the meaning (plane of content) of paradigmatic formtypes (grammemes). This meaning is revealed in the process of establishing the nature of correlation of grammemes WITH EACH OTHER (in contradiction to an analysis of isolated units), which enables one to get rid of a subjective interpretation of grammatical semantics.

— Oppositions are based on the principle of the contrast of language units; the very fact of contrast, in its turn, illustrates systematics and symmetry in the organization of linguistic signs.

— Oppositions are the method of the maximumly objective analysis of the semantic side of a bilateral (morphological) sign, since they are based on the principle of content correlation, at the same time reflecting the essence of relations among morphological units, i.e. to the greatest possible extent corresponding to their nature. Binarity is the simplest and most widely met kind of relations among the meanings of morphological units.

— Through oppositions we learn the grammatical meaning (specific for every particular language) of morphological forms as means of expressing grammatical categories (inherent for the given language), as well as the correlation of form and content and their interaction (but not identity). The members of oppositions are carriers of the minutest components of grammatical meaning — *DF*, being 'clusters' of these *DF*. Hence, oppositions are symbols of relations in the morphological system.

— While investigating the origin of the notion and method of oppositions and reasons for their appearance, one may conclude that the general trend in linguistic research of the twentieth century, which put on the agenda the cognition of systems and subsystems of language units, brought about the need to look for characteristics of ties and relations among different units of the language and their combinations. The best standard of measure of these ties are oppositions.

— Any consideration of the content of grammatical forms by the method of oppositions implies the introduction of a number of attendant notions (events and units), such as the notion of an unmarked and marked member, the invariant and its *DF*, the zero, the basis for comparison, neutralization of oppositions, and some others.

— The invariant points to relevant distinctions in a system, in the paradigmatics of the given sign. It is the *DF* that enter regular relations in oppositions. The *DF* (semes, morphosemes), necessarily imply correlations and define the place of the grammeme and its grammatical content in oppositions and within the system. Semes are distinguished by hierarchy, as are the members of oppositions. They are not homogeneous or equivalent, since the system in general lacks units with equal communicative functions. To define the composition of an opposition, those semes which are common for the pair of members of the opposition and those which distinguish them are singled out, since a purely external, morphemic analysis presents no data on the components of the plane of content of a morphological unit. Singling out oppositions allows the perception only of the paradigmatic meanings of grammemes, and not of syntagmatically conditioned ones caused by units of the environment.

— To characterize the members of oppositions it is important to distinguish the zero of the plane of expression and the zero of the plane of content.

In accordance with the nature of relations among morphological units in the plane of their content, mainly privative and equipollent oppositions are active in them. This fact testifies to a small variation in the nature of grammatical ties. Morphological units may be members of both intraparadigmatic and interparadigmatic oppositions; in both of them the members form a hierarchy.

— The connecting link between paradigmatics and syntagmatics is neutralization of members of oppositions, which introduces corrections into the systemic content of grammemes and illustrates the contextually conditioned contraction of paradigmatic sets. Neutralization of oppositions, as a syntagmatic, occasional, and distributionally conditioned event, radically differs from synonymy or variation which are phenomena constantly fixed in the language. In position of neutralization appears the unmarked or weaker member of the opposition, having a more general meaning compared with the marked, stronger member, Usually one component of the meaning of a grammeme is neutralized. Mainly privative oppositions are neutralizable, since the reduction of members with relevant *DF* (as in equipollent oppositions) is unnatural.

— The marked members of privative oppositions may be a source of redundancy in the system, since they may be easily removed from the system due to neutralization.

— With the definition of oppositions as intraparadigmatic or interparadigmatic is closely linked the definition of the notions of micro- and macrosystems, which are correlative and relative, and also correlate with the more general notion of the system. The system, in its turn, is closely linked with the structure. Microsystems are detected on the basis of compatibility–imcompatibility of *DF* (semes). The minimal distinction of different microsystems joined into one general macrosystem is one relevant *DF*. Thus, the units of the language are not only joined into one unique system, but into a number of micro- and macrosystems, the number of which depends upon the structural peculiarities of the given language.

— Applying the oppositional analysis to concrete material, the tense-aspect *MS* of the English verb enabled us to single out in it the invariants of sets and formtypes in their correlations with one another, on the basis of correlations of *DF* — carriers of grammatical meanings of the most abstract grammatical categories relevant for oppositions. Relevant categories are tense, temporal relativity (perfectness), and aspect (duration). Syntactically conditioned categories of person-number are irrelevant for oppositions among sets of verbal forms.

— Proceeding from the regular nature of correlation of morphosemes within the established oppositions, inconsistency is eliminated in the interpretation of grammatical content of separate sets of tense-aspect grammemes. The nature of oppositions of the form-sets of the given microsystem is also shown, as well as their behavior in regard to neutralization, which confirms the existing correlation among the verbal sets. Every grammeme is analysed as a cluster of semes, from which stem oppositions, strong and weak semes are detected, and tables and schemes are given, demonstrating their interaction. Oppositions also allow to distinguish the category of tense, showing different temporal axes of orientation of the action, from perfectness, marking the intervals among the actions on one axis.

— All these and other types of correlations in the tense-aspect *MS*, the interdependence of each type revealed in the process of analysing oppositions in action, are shown in chapters 10-14, and they present in a new light the relations repeatedly described in grammar.[1] The new method of analysis brings about new results, and allows for the comprehension in a new way of the relations in the system. In this connection it would be appropriate to mention the words of the reviewer of the book *A Functional View of Language* by A. Martinet, who, considering the book, wrote that it offers a ... "way of looking at old facts from a new angle."[2]

[1] All the notions introduced in this book, as well as the method of analysis and its results, are applicable to all the microsystems of the English (and not only English) verb which have not been analysed here.
[2] Mouton (1963) 242.

BIBLIOGRAPHY[1]

Admoni, V. G.
1964 *Osnovy teoriji grammatiki* (Moscow — Leningrad: Nauka).

Aiken, J. R.
1933 *A New Plan of English Grammar* (New York: Holt).

Akhmanova, O. S.
1961 "K voprosu ob osnovnyx ponjatijax metajazyka lingvistiki", *Voprosy jazykoznanija* 5, 115-121.

Allen, R. L.
1966 *The Verb System of Present-Day American English* (*Janua linguarum*, series practica 24) (The Hague: Mouton).

Barkhudarov, L. L. and D. A. Steling
1960 *Grammatika anglisjskogo jazyka* (Moscow: Literatura na inostrannyx jazykax).

Bazell, C. E.
1945-1949 "On Some Asymmetries of the Linguistic System", *Acta Linguistica* (Revue internationale de linguistique structurale) V. 3. (Copenhagen: Munksgaard), 139-145.
1949 "On the Neutralization of Syntactic Oppositions", *Travaux du cercle linguistique de Copenhague* 5 *(Recherches structurales)* (Copenhagen: Nordisk Sprog- og Kulturforlag), 77-86.
1966 "The Sememe", *Readings in Linguistics* II, ed. by E. Hamp, F. Householder, B. Austerlitz (Chicago-London: University of Chicago press), 329-340.

Bloch, B. and G. L. Trager
1942 *Outline of Linguistic Analysis (Special publications of the Linguistic Society of America)* (Baltimore: Waverley press).

Bodelsen, C. A.
1936-1937 "The Expanded Tenses in Modern English. An attempt at an explanation", *Englische Studien*, 71, 2 (Leipzig; Reisland), 220-238.

[1] Bibliography contains only those works which have been referred to in the book.

Bogdanov, V. V.
1965 "Opyt klassifikaciji minimal'nyx sintaksičeskix segmentov", *Voprosy obščego jazykoznanija* (Leningrad: University press), 3-25.
Brondal, Viggo
1943 *Essai de linguistique générale* (Copenhague: Munksgaard).
Brunot, F.
1953 *La pensée et la langue. Méthode, principes et plan d'une théorie nouvelle du langage appliquée au français* (Paris: Masson).
Brusendorf, A.
1930 "The Relative Aspect of the Verb in English", *A Grammatical Miscellany offered to O. Jespersen on his 70th birthday* (Copenhagen: Levin & Munksgard; London: George Allen & Unwin) 225-248.
Bryan, W. F.
1936 "The Preterit and the Perfect Tense in Present-Day English", *Journal of English and Germanic Philology* XXXV, 3 (Urbana, Illinois: The University of Illinois) 363-382.
Budagov, R. A.
1965 *Vvedenije v nauku o jazyke* (Moscow: Prosveščenije).
Bull, William E.
1960 *Time, Tense and the Verb. A study in theoretical and applied linguistics, with particular attention to Spanish (University of California publications in linguistics*, 19) (Berkeley & Los Angeles: University of California Press).
Buyssens, E.
1949 "Mise au point de quelques notions fondamentales de la phonologie", *Cahiers Ferdinand de Saussure* 8 (Genève: Georg) 37-60.
Callahan, J. J.
1938 *Science of Language* I (Pittsburg: Duquesne University).
Calver, E.
1946 "The Uses of the Present Tense Forms in English", *Language* 22, 4, 317-325.
Camara, J. Mattoso
1964 "Une categorie verbale: le futur du passé". *Proceedings of the Ninth International Congress of Linguists, Cambridge, Mass., August* 27-31, 1962, ed. by Horace G. Lunt (The Hague: Mouton) 547-551.
Cantineau, J.
1952 "Oppositions significatives", *Cahiers Ferdinand de Saussure* 10 (Genève: Georg) 11-40.
1955 "Le classement logique des oppositions", *Word* 11, 1, 1-9.
Curme, G. O.
1935 *A Grammar of the English Language* II, *Parts of Speech and Accidence* (New York, etc.: Heath).
Daneš, F. and J. Vachek
1964 "Prague Studies in Structural Grammar Today", *Travaux linguistiques de Prague* 1 (L'école de Prague d'aujourd'hui) (Prague: l'Académie tchécoslovaque des sciences) 21-31.
Dietrich, G.
1955 *Erweiterte Form. Präteritum und Perfektum im Englischen. Eine*

Aspekt- und Tempusstudie (*Studien und Texte zu englischen Philologie*, ed. by Josef Raith, 2) (München: Hueber).

Diver, W.
1963 "The Chronological System of the English Verb", *Word*, 19, 1, 141-181.

Firth, J. K.
1957 "A Synopsis of Linguistic Theory", *Studies in Linguistic Analysis* (Special volume of the Philological Society) (Oxford: Basil Blackwell) 1-32.

Fortunatov, F. F.
1956 *Izbrannyje trudy*, V. I (Moscow: Učpedgiz).

Francis, William N.
1958 *The Structure of American English* (New York: Ronald).

Frei, Henri
1929 *La grammaire des fautes* (Paris: Geuthner).
1950 "Zéro, vide et intermittent", *Zeitschrift für phonetik und allgemeine Sprachwissenschaft*, 3/4, ed. by Diedrich Westermann (Berlin: Akademie-Verlag GMBH) 161-191.

Garey, H. B.
1955 "The Historical Development of Tenses from Late Latin to Old French", *Language Supplement* 31, I(II), (*Language dissertation* 51).
1957 "Verbal Aspect in French", *Language* 33, 2, 91-110.

Glinz, H.
1952 *Die innere Form des Deutschen. Eine neue deutsche Grammatik* (Bern: Francke).

Godel, R.
1953 "La question des signes zéro", *Cahiers Ferdinand de Saussure* 11 (Genève: Librairie E. Droz), 31-41.

Gougenheim, G.
1938 *Système grammatical de la langue française* (Paris: Ponteix-d'Artrey).

Groot, A. W. de
1939 "Les oppositions dans les systèmes de la syntaxe et des cas", *Mélanges de linguistique offerts à Ch. Bally* (Genève: Georg) 107-127.

Guillaume, G.
1929 *Temps et verbe. Théorie des aspects, des modes et des temps* (Paris: Champion).

Haas, W.
1957 "Zero in Linguistic Description". *Studies in Linguistic Analysis* (Special volume of the Philological Society) (Oxford: Basil Blackwell) 33-53.

Halliday, M. A. K.
1957 "Some Aspects of Systematic Description and Comparison in Grammatical Analysis", *Studies in Linguistic Analysis* (Special volume of the Philological Society) (Oxford: Basil Blackwell) 54-67.

Harris, Zellig
1951 *Methods in Structural Linguistics* (Chicago: University of Chicago press).

Hartmann, Peter

1962 *Grammatik und Grammatizität* (*Janua Linguarum*, series major 34) (The Hague: Mouton).

Hatcher, Anna G.

1951 "The Use of the Progressive Form in English. A new Approach", *Language* 27, 3, 254-280.

Hjelmslev, Louis

1928 *Principes de grammaire générale* (København: Host).

1939 "Note sur les oppositions supprimables", *Travaux du Cercle linguistique de Prague* 8 (Études phonologiques dédiées à la mémoire de N. S. Trubetzkoy) (Prague: Jednota českých matematikû A fysikû) 51-57.

1968 *Prolégomènes a une théorie du langage (Collection "Arguments")* (Paris: Les éditions de Minuut).

Holk, André G. F. van

1958 *On the Semantic mechanism of the Russian Tenses* (Dutch Contributions to the Fourth International Congress of Slavicists, Moscow, September, 1958, Preprint) (The Hague: Mouton).

Horálek, K.

1964 "A propos de la théorie des oppositions binaires", *Proceedings of the Ninth International Congress of Linguists, Cambridge, Mass., August* 27-31, 1962, ed. by Horace G. Lunt (The Hague: Mouton) 414-117.

Ilyish, B. A.

1965 *The Structure of Modern English* (Moscow-Leningrad: Prosveščenije).

Irtenjeva, N. F.

1956 *Grammatika sovremennogo anglijskogo jazyka* (Moscow: Ministerstvo prosveščenija RSFSR).

Isačenko, A. V.

1960 *Grammatičeskij stroj russkogo jazyka v sopostavleniji s slovackim* (Bratislava: Slovackaja Akademija nauk).

1961 "O grammatičeskom značeniji", *Voprosy jazykoznanija* 1, 28-43.

1963 "Binarnost', privativnyje oppoziciji i grammatičeskije značenija", *Voprosy jazykoznanija* 2, 39-56.

Ivanov, V. V. and V. N. Toporov

1960 *Sanskrit* (Moscow: Vostočnaja Literatura).

Ivanova, I. P.

1961 *Vid i vremja v sovremennom anglijskom jazyke* (Leningrad: University press).

Jakobson, Roman

1932 "Zur Struktur des russischen Verbums", *Charisterio Guilelmo Mathesio quinquagenario, A discipulis et Circuli linguistici Prahensis sodalibus oblata* (Prague Sumptibus "Pražsky linguistický kroužek") 74-84.

1936 "Beitrag zur allgemeinen Kasuslehre", *Travaux du Cercle linguistique de Prague* 6 (Études dédiées au quatrième Congrès de linguistes) (Prague) 240-288.

1939 "Signe zéro", *Mélanges de linguistique offerts à Charles Bally* (Genève: Georg) 143-152.

1957 *Shifters, Verbal Categories and the Russian Verb* (Harvard University

Department of Slavic languages and literatures. Russian language project) (Cambridge, Mass.: Harvard University press).
1958 *Morfologičeskije nabljudenija nad slavjanskim sklonenijem* (American contributions to the Fourth International Congress of Slavists, Moscow, Sept., 1958, Preprint) (The Hague: Mouton).
1959 "Boas' View of Grammatical Meaning", *The American Anthropologist* (The Anthropology of Franz Boas. Essays on the centennial of his birth), ed. by Walter Goldschmidt, The American anthropological association 61, 5, part 2, Memoir 89, October 1959, 139-145.

Jakobson, R., C. Gunnar, G. M. Fant and M. Halle
1955 *Preliminaries to Speech Analysis. The Distinctive Features and their Correlates* (Technical report 13), 2nd printing (Boston: Acoustics Laboratory Massachusetts Institute of Technology).

Jakobson, Roman and Morris Halle
1956 *Fundamentals of Language* (*Janua linguarum* 1) (The Hague: Mouton).

Jarceva, V. N.
1963 "Ob analitičeskix formax slova, *Morfologičeskaja struktura slova v jazykax različnyx tipov* (Moscow-Leningrad: Academy of Sciences) 52-60.

Jespersen, Otto
1951 *The Philosophy of Grammar* (London: George Allen & Unwin).

Joos, Martin
1964 *The English Verb, Form and Meanings* (Madison-Milwaukee: University of Wisconsin press).

Juilland, Alphonse G.
1961 *Outline of a General Theory of Structural relations* (*Janua linguarum* 15) (The Hague: Mouton).

Kahn, F.
1954 *Le système des temps de l'indicatif chez un Parisien et chez une Bâloise* (Genève: Droz).

Kantor, J. R.
1936 *An Objective Psychology of Grammar* (Bloomington: Indiana University press).

Karcevskij, S.
1927 *Système du verbe russe* (Prague: Plamja).
1964 "Du dualisme asymétrique du signe linguistique", *A Prague School Reader in Linguistics* (Indiana University Studies in the History and Theory of Linguistics), ed. by Joseph Vachek (Bloomington: Indiana University press) 81-87.

Khlebnikova, I. B.
1963 "Funkcional'nyje osnovy vydelenija klassov slov", *Problemy morfologičeskogo stroja germanskix jazykov* (Moscow: Academy of sciences) 36-47.
1964a "Binary Relations in Morphology", *Philologica Pragensia* 2 (Prague: Academy of sciences) 150-158.
1964b "O neitralizaciji oppozicij v morfologiji", *Inostrannyje jazyki v vyššei škole.* (Moscow: Vyššaja škola) 54-64.
1965a "O granicax morfologiji i sintaksisa", *Filologičeskije nauki* 4, 124-132.

1965b "Homonymy and the Subjunctive Mood in English", *Philologica Pragensia* 2-3 (Prague: Academy of sciences) 238-245.
1966a "Nekotoryje voprosy tipologičeskogo sopostavlenija vido-vremennyx sistem", *Strukturno-tipologičeskoje opisanije sovremennyx germanskix jazykov* (Moscow: Nauka) 62-73.
1966b "Zametki ob analitičeskoj forme", *Učenyje zapiski Orekhovo-Zuevskogo pedagogičeskogo instituta* 27, 1, 44-50.
1967 "Notes on Grammatical Category", *Philologica Pragensia* 1 (Prague: Academy of sciences) 28-33.
1970 "Sinonimija i variantnost' morfologičeskix jedinic", *Voprosy anglijskoj filologiji* (Učenyje zapiski fakul'teta inostrannyx jazykov, 3) (Tula: Pedagogičeskij Institut) 3-23.
1971 *Soslagatel'noje naklonenije v anglijskom jazyke (kak obščelingvističeskaja problema)* (Kalinin: Miniterstvo Prosveščenija RSFSR).

Kurylowicz, J.
1949a "Na notion de l'isomorphisme" *Esquisses linguistiques* (Wroclav-Krakow: Academy of sciences, 1960) 16-26.
1949b "Linguistique et théorie du signe", *Esquisses linguistiques* (Wroclav-Krakow: Academy of sciences, 1960) 7-15.
1953 "Aspect et temps dans l'histoire du persan", *Esquisses linguistiques* (Wroclav-Krakow: Academy of sciences, 1960) 109-118.
1956 *L'apophonie en indo-européen* (Wroclav: Academy of sciences).

Kuznecov, P. S.
1958 "O differencial'nyx priznakax fonem", *Voprosy jazykoznanija* 1, 55-61.

Leška, O.
1964 "Zur Invariantenforschung in der Sprachwissenschaft", *Travaux linguistiques de Prague* 1 (L'école de Prague d'aujourd'hui) (Prague: L'Academie tchécoslovaque des sciences) 85-93.

Lomtev, T. P.
1958 *Osnovy sintaksisa sovremennogo russkogo jazyka* (Moscow: Učpedgiz).

Makaev, E. A.
1962 "Ponjatije davlenija sistemy i ijerarxija jazykovyx jedinic", *Voprosy jazykoznanija* 5, 47-52.

Malmberg, B.
1963 *Structural Linguistics and Human Communication* (Berlin: Springer).

Marchand, H.
1955 "On a Question of Aspect: a comparison between the progressive form in English and that in Italian and Spanish", *Studia Linguistica* (Revue de linguistique générale et comparée) 9, 1 (Lund: Gleerup; Copenhagen: Munksgaard) 45-52.

Marckwardt, A. H.
1942 *Introduction to the English Language* (New York, Toronto: Oxford University press).

Marcus, S.
1962 "Teorija grafov, lingvističeskije oppoziciji i invariantnaja struktura", *Problemy strukturnoj lingvistiki* (Moscow: Nauka) 22-30.
1963 "Logičeskij aspekt lingvističeskix oppozicij", *Problemy strukturnoj lingvistiki* (Moscow: Nauka) 47-74.

Martinet, André

1936 "Neutralisation et archiphonème", *Travaux du Cercle linguistique de Prague* 6 (Études dédiées au quatrième Congrès de linguistes) (Prague) 46-57.

1949 "About Structural Sketches", *Word* 5, 1, 13-35.

1953 "Structural Linguistics", *Anthropology Today* (An Encyclopedic Inventory), ed. A. L. Kroeber (Chicago, Illinois: The University of Chicago Press) 574-586.

1960 *Éléments de linguistique générale*, (*Collection Armand Colin*, 349, Section de littérature) (Paris: Librarie Armand Colin).

1962 *A Functional View of Language* (Oxford: Clarendon press).

1964 "Structural Variation in Language", *Proceedings of the Ninth International Congress of Linguists, Cambridge, Mass., August* 27-31, 1962, ed. by Horace Lunt (The Hague: Mouton) 521-532.

Meier, G. F.

1961 *Das Zéro-Problem in der Linguistik* (Berlin: Akademie Verlag).

Møller, K.

1949 "Contribution to the Discussion Concerning 'Langue' and 'Parole'", *Travaux du Cercle linguistique de Copenhague* 5 (*Recherches structurales)* (Copenhagen: Nordisk Sprog og Kulturvorlag) 87-96.

Moulton, W. G.

1963 Review of *A Functional View of Language* by A. Martinet, *Word* 19, 2, 242-245.

Mukhin, A. M.

1964 *Funkcional'nyj analiz sintaksičeskix elementov* (Moscow-Leningrad: Nauka).

Notion

1957 *Notion de structure et structure de la connaissance* (Centre International de synthèse, XX-me semaine de synthèse, 18-27 avril 1956), ed. by Albin Michel (Paris: Albin Michel).

Ogden, Charles K.

1932 *Opposition. A Linguistic and Psychological Analysis (Psyche Miniatures)* (London: L. Paul, Trench Trubner & Co).

Palmer, F. R.

1968 *A Linguistic Study of the English Verb* (*Miami Linguistic series* No. 2) (Coral Gables, Florida: University of Miami press).

Peškovskij, A. M.

1956 *Russkij sintaksis v naučnom osveščeniji* (Moscow: Učpedgiz).

Pike, K. L.

1954-1960 *Language in Relation to a Unified Theory of Structure of Human Behavior*. Part I — 1954; Part II — 1955; Part III — 1960 (Glendale, California: Summer Institute of Linguistics).

1958 "On Tagmemes, Née Grammemes", *International Journal of American Linguistics* 24, 4, 273-278.

Pospelov, N. S.

1955 "Prjamoe i otnositel'noje upotreblenije form nastojaščego i buduščego vremeni glagola v sovremennom russkom jazyke", *Issledovanija po grammatike sovremennogo russkogo literaturnogo jazyka* (Moscow:

Academy of sciences) 206-146.
Pottier, B.
1962 *Systématique des éléments de relation* (Paris: Librairie Klincksieck).
Prieto, L. S.
1954 "Traits oppositionnels et traits contrastifs", *Word* 10, 1, 43-59.
1964 *Principes de noologie* (*Janua linguarum*, series minor 35) (The Hague: Mouton).
Reformatskij, A. A.
1957 Čto takoje strukturalism?", *Voprosy jazykoznanija* 6, 25-37.
1961 "Dixotomičeskaja klassifikacija differencial'nyx priznakov i fonematičeskaja model' jazyka", *Voprosy teoriji jazyka v sovremennoj zarubežnoj lingvistike* (Moscow: Academy of sciences), 106-122.
Reichenbach, Hans
1947 *Elements of Symbolic Logic* (New York: Macmillan).
Revzin, I. I.
1960 "O sil'nyx i slabyx protivopostavlenijax v sisteme padežej sovremennogo nemeckogo jazyka", *Voprosy jazykoznanija* 3, 82-85.
1964 "K logičeskomu obosnovaniju teoriji fonologičeskix priznakov", *Voprosy jazykoznanija* 5, 59-65.
Roberts, P.
1954 *Understanding Grammar* (New York: Harper).
Ruipérez, M. S.
1953 "The Neutralization of Morphological Oppositions as Illustrated by the Neutral Aspect of the Present Indicative in Classical Greek", *Word* 9, 3, 241-252.
Russell, Bertrand
1948 *Human Knowledge. Its Scope and Limits* (New York: Simon & Schuster).
Šabršula, J.
1966 "Un problème de la périphérie du système morphologique: a propos des formations prémorphologiques", *Travaux linguistiques de Prague* 1 (L'école de Prague d'aujourd'hui) (Prague: l'Académie Tchécoslovaque des sciences), 183-192.
Šaumjan, S. K.
1962 *Problemy teoretičeskoj fonologiji* (Moscow: Academy of sciences).
1965 *Strukturnaja lingvistika* (Moscow: Nauka).
Saussure, Ferdinand de
1915 *Cours de linguistique générale* (Paris: Payot, 1949[4]).
Ščerbak, A. M.
1965 "Lingvističeskije zametki", *Voprosy jazykoznanija* 6, 16-23.
Sčur, G. S.
1964 "Some Remarks Concerning the Germanic Future", *Transactions of the Philological Society* (Hertford: Austin), 49-57.
Skalička, V.
1935 *Zur Ungarischen Grammatik* (Prague: Charles University).
Slama-Cazacu, Tatiana
1961 *Langage et contexte. Le problème du langage dans la conception de l'expression et de l'interpretation par des organisations contextuelles* (*Janua linguarum*, series major, 6) (The Hague: Mouton).

Smirnickij, A. I.
1956 "Analitičeskije formy", *Voprosy jazykoznanija* 2, 41-52.
1957 *Sintaksis anglijskogo jazyka* (Moscow: Literatura na inostrannyx jazykax).
1959 *Morfologija anglijskogo jazyka* (Moscow: Literatura na inostrannyx jazykax).
Spang-Hanssen, H.
1959 *Probability and Structural Classification in Linguistic Description* (Copenhagen: Rosenkilde & Bagger).
1964 "Mathematical Linguistics — A Trend in Name or in Fact?", *Proceedings of the Ninth International Congress of Linguists, Cambridge, Mass., August* 27-31, 1962, ed. by Horace G. Lunt (The Hague: Mouton) 61-66.
Stankiewicz, E.
1961 "Grammatical Neutralization in Slavic Expressive Forms", *Word* 17, 2, 128-145.
Sten, H.
1952 *Les temps du verbe fini (indicatif) en français moderne* (København: Munksgaard).
Stepanov, J. S.
1966 *Osnovy jazykoznanija* (Moscow: Prosveščenije).
Strang, Barbara
1968 *Modern English Structure* (London: Edward Arnold).
Sunik, O. P.
1966 *Obščaja teorija častej reči* (Moscow-Leningrad; Nauka).
Sweet, Henry
1892 *A New English Grammar, Logical and Historical* (Oxford: The Clarendon press, 1900).
Tellier, A.
1962 *Les verbes perfecto-presents et les auxiliaires de mode en Anglais ancien* (VIII-e s.—XVI-e s.) (Paris: Librairie Klincksieck).
Tesniére, L.
1939 "Théorie structurale des temps composés", *Mélanges de linguistique offerts à Ch. Bally* (Genève: Georg), 153-83.
Togeby, K.
1953 *Mode, aspect et temps en Espagnol* (*Det Kongelige Danske Videnskabernes selskab Historisk-filologiske Medelelser*, bind 34, N 1) (København: Munksgaard).
Travaux
1957 *Travaux de l'institut de linguistique* 2. *La notion de neutralisation dans la morphologie et le lexique* (Faculté des lettres de l'Université de Paris) (Paris: Librairie C. Klincksieck).
Travaux
1964-1968 *Travaux Linguistiques de Prague* (L'école de Prague d'aujourd'hui) 1, 2, 3 (1964, 1966, 1968) (Prague: l'Académie Tchécoslovaque des sciences).
Trnka, B.
1960 "Autonomous and Syntagmatic Words", *Studii si cercetari lingvistice.*

Omagiu Lui Al. Graur cu prilejul impliniri A 60 de ani 3. (Bucuresti: Akademiei Republicii) 761-763.
1961 "Principles of Morphological Analysis", *Philologica* 2, 3, 129-137.
1966 "On the Basic Categories of Syntagmatic Morphology", *Travaux Linguistiques de Prague* 2 (L'école de Prague d'aujourd'hui) (Prague: l'Académie Tchécoslovaque des sciences), 165-169.

Trost, P.
1939a "Zum lateinischen Konditionalsatz", *Glotta* XXVI (Zeitschrift für griechische und lateinische Sprache, ed. Paul Kretschmer) (Göttingen: Vanderhoeck & Ruprecht) 207-11.
1939b "Bemerkungen zum deutschen Vokalsystem", *Travaux du Cercle linguistique de Prague* 8 (Études phonologiques dédiées à la mémoire de N. S. Trubetzkoy) (Prague: Jednota českych matematikû A fisikû) 319-326.

Trubetzkoy, N. S.
1934 *Das morphonologische System der russischen Sprache* (*Travaux du Cercle linguistique de Prague* 5^2. Description phonologique du russe moderne, part 2) (Prague: Jednota československých matematikû A fysikû; Leipzig: Otto Harrassowitz).
1936a "Die Aufhebung der phonologischen Gegensätze", *Travaux du Cercle linguistique de Prague* 6 (Études dédiées au quatrième Congres de linguistes) (Prague) 29-45.
1936b "Essai d'une théorie des oppositions phonologiques", *Journal de psychologie normale et pathologique* XXXIII, 1, 2, 5-18.
1939 *Grundzüge der Phonologie* (*Travaux du Cercle linguistique de Prague* 7) Prague).

Twaddell, W. F.
1960 *The English Verb Auxiliaries* (Providence, Rhode Island: Brown University press).

Volockaya, Z. M., T. N. Mološnaja and T. M. Nikolajeva
1964 *Opyt opisanija russkogo jasyka v jego pis'mennoj forme* (Moscow: Nauka).

Voronzova, G. N.
1960 *Očerki po grammatike anglijskogo jazyka* (Moscow: Literatura na inostrannyx jazykax).

Weber, H.
1954 *Das Tempussystem des Deutschen und des Französischen. Übersetzungs und Strukturprobleme* (*Romanica Helvetica,* 45) (Bern: Francke).

Zawadowski, L.
1959 *Constructions grammaticales et formes periphrastiques* (*Prace Jezykoznawcze* 18) (Kraków: Akad. nauk).

Zinder, L. R.
1962 "O protivopostavlenijax v sisteme jazyka", *Vestnik* 20 (Leningrad: University Press), 121-124.

Zveginzev, V. A.
1962 *Očerki po obščemu jazykoznaniju* (Moscow: University press).

Žirmunskij, V. M.
1961 "O granicax slova", *Voprosy jazykoznanija* 3, 3-21.

Bei Fragen zur Produktsicherheit wenden Sie sich bitte an:
If you have any questions regarding product safety,
please contact:

Walter de Gruyter GmbH
Genthiner Straße 13
10785 Berlin
productsafety@degruyterbrill.com